ORGANIC FARMING: METHODS and MARKETS

An Introduction to Ecological Agriculture

Edited by

Robert Steffen

Floyd Allen

James Foote

of Organic Gardening and Farming®

Rodale Press, Inc.,
Book Division
Emmaus, Pa. 18049

The text was written especially for this book and is based on material which has appeared in Organic Gardening and Farming magazine.

Standard Book Number 0-87857-019-5
Library of Congress Card Number 76-190200

Printed in the U.S.A.
on recycled paper

B-241

First Printing — June 1972

CONTENTS

PART I TODAY'S FARMING

Chapter

PART II ORGANIC FARMING METHODS

Chapter

PART III ORGANIC FOOD MARKETING

Chapter

PART IV ORGANIC FARM DIRECTORY

Chapter

Enriching The Earth

To enrich the earth I have sowed clover and grass to
grow and die. I have plowed in the seeds of winter
grains and of various legumes, their growth to be
plowed in to enrich the earth. I have stirred into the
ground the offal and the decay of the growth of past
seasons and so mended the earth and made its yield
increase. All this serves the dark. Against the shadow
of veiled possibility my workdays stand in a most
asking light. I am slowly falling into the fund of things.
And yet to serve the earth, not knowing what I serve,
gives a wideness and a delight to the air, and my days
do not wholly pass. It is the mind's service, for when
the will fails so do the hands and one lives at the
expense of life. After death, willing or not, the body
serves, entering the earth. And so what was heaviest
and most mute is at last raised up into song.

—Wendell Berry

PART I

TODAY'S FARMING

CHAPTER 1

The Need For Ecological Agriculture

AN ESSENTIAL PART of the pioneer spirit of America was to conquer and subdue the wilderness. It is not hard to understand how such an attitude developed. North America was a cruel vast land seemingly without limit in space and natural resources. This abundance of course was what induced men to leave the old world and endure the hardships of the new wilderness. Conservation was not uppermost in the minds of the early pioneers. What the immigrants brought with them in the way of skills in good husbandry and conservation were soon forgotten or simply ignored because there was so much of everything, land, water, forests and minerals.

But the frontier is gone and we finally have come to the realization that the earth is limited in its capacity to produce and in the natural resources we have taken for granted for these two hundred some years.

Mankind must reexamine its relationship with the planet Earth. This reevaluation must be undertaken by all men in all areas of human activity. This, of course, involves agriculture because farming is man's principal activity which has direct contact with nature. In this reevaluation man must improve his relationship with the soil and in doing this he will find that it becomes easier for all humankind to improve their relationship with each other.

As a farmer, how do you proceed with this reevaluation? How are you involved in the current discussion of environmental problems? How can my way of farming affect the environment? Why should I be interested in ecological agriculture?

Try to look at your environment as part of a living organism. What makes up the anatomy of this organism? What makes it alive? What makes it bloom? What makes it die? You are part of a world made up of many living organisms. Everything you do will in some way affect not only yourself, your family, your farm, but your neighborhood, your watershed, your country and your planet. Realization of this will help to appreciate infinitely more the natural world about you.

Ecologists will tell you that everything is connected to everything else. This is a bit hard to realize until you think about it a while. I would urge you to think about this often, as you go about your work in raising crops and feeding livestock. You are in a better position to understand this than most people living in an increasingly urbanized society. The need to understand this principle of ecology really is critical and it will require a great amount of time and human effort to educate the masses about its significance and to carry out the needed changes that are involved. The better the general public understands this, the easier it will be for you to get done what you need to do on your farm, to make it an ecologically-oriented operation. But the obstacles facing urban society and its educational processes are many and formidable. You can't really wait for this to take place. It may be too late.

The next time you get to your favorite resting place on your farm, perhaps a favorite tree on a hill overlooking your own land and that of your neighbors, shut off the engine of your machine and pause for a moment. I hope it is quiet. It

1

is so hard, in fact almost impossible for many of us to actually find a quiet place, even in the country. But look about you and be thankful for just being there. Think of the millions of human beings struggling in the dead or dying environs of the urban wilderness. Instead of stopping to look around and enjoying for a moment, such a panorama of living things such as you can enjoy, your urban friend hurries to nowhere, through an artificial world and hears nothing because of the noise. For reasons he cannot understand, he has become entrapped in a system that is ever growing and threatens to engulf us all.

Somehow you must not allow yourself to be swallowed by this insatiable monster that really seems wholly out of control and encompasses ever more and more and more people.

I realize this is almost like asking the impossible. I am familiar with the lure that hangs before you and your family, especially the young people. Our materialistic philosophy that accompanies our technology is not giving us the proper values. Economic return for your efforts are meager. There are those who say that our technological society has within it the seeds of its own destruction. Many people are convinced that the system must grow or it will die. The question we must ask is how big can it get before it consumes itself?

Somehow you must retain some independence, some ability to live by your wits; your own strength; your own ingenuity; your own skills, because in doing this you do help mankind.

Treat your farm as a living organism. Strive to make it a more self-renewing system so that in an emergency it can indeed live by itself. Treat your farming in such a way that it will go on producing for the next century, not just for your lifetime. This again is easier said than done. But we hope that our society will see the need for this soon enough and make it possible to do what needs to be done to assure future generations of a strong healthy and viable rural community.

Perhaps you should give your farm a name. After all it is alive. It has a personality. It has character. You will love it more and take better care of it. Maybe your children will also become more interested. The need to keep young people on farms is urgent. Young people are not satisfied with modern society's values. Farming needs a new image. We need to look upon it once more as a way of life. It needs new direction and a new sense of urgency that the young people can give it. It is ironic that there are thousands of young people in the cities who see the countryside as their only hope for a meaningful life, while the rural youth are moving away from the country.

The future will see more and more urban expansion in spite of our efforts and awareness of what is happening. But this movement of the cities into the country can indeed be a mutually satisfactory development. This blending of the rural and urban landscape can be an economically and ecologically sound development as long as society understands fully the need to preserve our natural resources and the quality of our environment. This development is going to require a tremendous change in attitudes toward mutual problems and certainly a change in society's priorities.

But there are many signs today indicating that this process has already begun. The many environmental groups around the country will understand your situation. They know that these changes are inevitable and are even now beginning to move in this direction. In developing your marketing programs with these urban, or off-farm consumers, keep this in mind. Society must understand that it is possible to produce an abundant food supply and still maintain a liveable environment if they cooperate. This educational process is perhaps your responsibility more than anyone else's.

There is an ever-growing group of scientists who see the need for changing our ways of farming, and in fact the need to change our whole way of living. The question is, can you survive in the interim? We hope this book will help you in at least a small way.— *Robert Steffen*

2

CHAPTER 2

What's Your Future In Farming?

"AGRICULTURE IS RAPIDLY becoming a high-investment, low-wage industry. This is unhealthy!"

That statement by Dr. Marion Clawson, who directs studies in land use and management for Resources for the Future, Inc. in Washington, D.C., illustrates another major weakness in current agricultural programs. Dr. Clawson documents his case for "A New Policy Direction for American Agriculture," in a paper by that title in the Jan.-Feb. 1970 issue of the *Journal of Soil and Water Conservation*.

Recalling when the Depression of the Thirties brought about foreclosed farms, unbelievably low price levels ("in some places a bushel of oats would not buy a postage stamp"), Dr. Clawson traces the government agricultural programs that instituted various forms of production control, price support and storage of surpluses. He writes:

"Through Democratic and Republican administrations for nearly 40 years, this trinity of production control (or supply management), price support, and surplus storage has remained central to all agricultural programs. . . . Through depression and boom, through war and peace, the same basic programs have continued."

Dr. Clawson maintains that these programs have largely failed, and even worse, have contributed largely to the collapse of our rural communities. Landowners rather than farmers have been the recipients of federal subsidies.

Continues Dr. Clawson:

"One major effect . . . has been a persistent increase in farm land prices. The total value of all farm real estate in the U.S. has more than doubled since 1954. In the same period, net income to agriculture as a whole has remained nearly constant."

We have created an artificial land scarcity through federal agricultural programs. At present, more than 50 million acres of cropland are held out of production by USDA farm programs.

Years ago, most economists thought that farmers would benefit by rising land prices. But it hasn't worked out that way. During the past generation, points out Dr. Clawson, average farm size in the U.S. has doubled. About half of all purchases of farmland have been by established farmers to increase the size of their farm. What they may have gained in increased prices of land they previously owned, they have lost in increased prices of land which they have bought.

"The total value of farm real estate in the U.S. today is so high that an interest return at competitive interest rates leaves nothing for wages to the farmer," declares Dr. Clawson.

"Federal agricultural programs of the past generation have given only incidental attention to the human problems of farm people . . . The small farmer, whose basic problem is inadequate farm size, or the farm laborer or tenant without even a minimum farm has benefited little or not at all from federal farm programs."

The last thing we've tried to do in this country has been to try to keep people

3

down on the farm. The net effect of mechanizing for lowest-cost food production has had a predictable effect. We've reduced the number of farm families from more than 8 million in 1935 to less than 4 million today. The general family diversified farm has largely disappeared. By 1968, three million farmers were producing enough food, feed, fiber and other raw materials for 200 million Americans.

According to Dr. Clawson, farm numbers have been cut in half, primarily because young men have been so repelled by the prospect of low incomes that they refuse to enter farming. "Farming has become an old man's occupation," he writes.

We've developed and practiced the philosophy that, since all food is equal regardless of how it is produced, the role of American agriculture (and USDA policy) is to produce food as cheaply and mechanically as possible. Just like we do with cars. Or envelopes. Or any other product.

But now a growing number of people—in Washington, in Emmaus, in New York, in San Francisco, in Front Royal, in Cambridge, in Butler, all over—are questioning that philosophy. Is all food equal in quality regardless of how it is grown? Can the *methods* of food production be adding to our pollution crisis, when such methods demand the use of pesticides and artificial nitrates, or the use of mercury compounds?

Must we allow such methods to continue to drive people off the land? Isn't there a solution that can mean more income as well as more sense to American farmers?

The greatest number of people asking those questions and doing something to solve them are readers of *Organic Gardening and Farming* magazine.

These people practice not only a "land ethic" for themselves, but also make up the most concentrated group of growers, suppliers, wholesalers, retailers and consumers of organically-grown foods and ecological products in America.

Together they are trying to create a new change in the philosophy of American agriculture. It's far more than the negative side of not using dangerous pesticides and polluting fertilizers. It's far more even than using natural fertilizers and soil-conditioners.

What they are trying to show is that there is a rapidly growing trend for food grown with more man-power. Sure, we know that eggs produced by chickens which are allowed to run around will cost more than eggs produced by chickens that are penned up four to a cage. But so what? Those organically-produced eggs will get and deserve a higher price. Part of that higher price will provide decent wages for more farm workers—right now those farm workers of the future may be sweating it out in some urban tenement.

We believe that the organic market can provide a distribution channel for foods that can mean higher prices to the grower.—*Jerome Goldstein*

CHAPTER 3

Is The Crisis In Agriculture Hurting You?

THERE HAS NEVER BEEN A CIVILIZATION that produced as much food with so little human energy. Our technology, together with abundant resources of land, water, a favorable climate, and abundant fossil fuels, make it possible. We have become experts in exploiting the available natural resources. Very little thought has been given to replenishing and/or recycling any wastes.

Agriculture has moved from what was substantially a self-renewing industry to one that is almost completely dependent on outside sources for its operation. As a result, farming has moved from a way of life to a factory-in-the-field concept. The effects of these changes on the biology of agriculture is, indeed, significant and far-reaching.

We point with pride to the fact that one man can now produce the food requirements for himself and about 30 and more other people. Although this may be true to a large extent, it is not altogether the whole story, because of the number of people and resources that are used in industry to supply this one farmer's needs. Again we are replacing human energy with the stored energy of the fossil fuels.

The machinery, inorganic fertilizers, and other products produced by industry for the farmer, are of course using up some of our natural resources. Again the stored fossil fuels which are not inexhaustible, the precious oxygen which scientists tell us we are using faster than it is being produced, and the various mineral stores are being depleted in this process. In many cases use of these products is obviously becoming a major source of soil and water pollution. It is quite generally known, for example, that almost all the streams and rivers in Illinois are carrying high levels of nitrates. The city of Bloomington, Illinois, gets its water from a lake now contaminated to such an extent that doctors recommend not using this water for babies. Contamination of wells in the Midwest—from organic wastes as well as inorganic fertilizers—is also well-known.

The insect pests that always result from a monoculture are becoming more of a problem every year. New and apparently more toxic materials must continually be used. The monoculture becomes attractive, of course, because of industrialized agriculture and, indeed, becomes a necessity because of the tremendous investment the farmer has in his machines.

Meanwhile, the livestock that has been for centuries an integral part of the traditional methods of agriculture, is also moving from the farms. Livestock, like people, have been concentrated in small areas. Again our drive for efficiency and our industrial concepts of producing a given product for the least amount of cost, is our only yardstick for measuring success. Disease is becoming a greater problem as a result of high concentrations of animals.

Waste disposal problems have in fact reached such proportions that even the urban dweller is well aware of them. Mountains of manure have become a real source of pollution, while back on the farm, where the substitute is used, it also has become a source of pollution. Again the nitrogen cycle, that every high-

5

school biology student learns about, has been broken. Soils are rapidly losing their nitrogen-fixing bacteria. Indeed, some have largely lost their ability to properly nourish the growing plant. They have become like the dope addict who needs his "shot" to function at all.

The industrialization of agriculture has so speeded up the biological functions of the soil that they simply cannot maintain a healthy environment for the growing plant. Trace mineral problems have become very common, largely because of the reduced microbial action in the soil. There is roughly a 50% loss of organic matter in our soils—so very necessary to maintain their tilth, water-absorbing capability, and active biological life. Only a healthy soil can produce a healthy plant. Only a healthy plant can survive in nature's domain, and our attempts to fight insects and fungi that attack our plants should be tempered by this fact, and serve notice for us to proceed with caution.

Agriculture will perhaps never again become the totally self-renewing industry that it once was. But if we are to observe the laws of biology and adhere to nature's law that all things must operate in a cycle, how are we going to accomplish this with 94% of the population living in the cities, and only 6% left on the farms? How can we recycle the tremendous concentrations of food which of course becomes "waste," when the centers of population are so far removed from the source of food? How can we recycle the tremendous amounts of waste from our feedlots and poultry houses when they are so far removed from the land?

Now we are getting to the real gut issue and the real test as to whether or not we really *can* or *will* obey the biological laws that we have been ignoring since the beginning of the industrial revolution, and especially since the beginning of modern agricultural practices. This is what I feel is the real crisis in agriculture —or, for that matter, the crisis for all of the so-called developed countries. We have become accustomed to an abundance of cheap food that is the result of our taking shortcuts in nature's biological cycle. We have accumulated a huge debt, and nature will have her due in one way or another. She is very patient, but very demanding. A century is but a moment in her web of life. If we choose not to pay the debt, she may have to wait until man disappears from the scene, when she will once again, in her own way, restore the earth as we found it. Remember the tremendous fertility we found in the North American continent was not the work of man.

So the question needs to be answered. Do we really want to start living as though we want to remain on this earth as a species? Do we really have the technology that will enable us to continue to ignore the biological laws? This is the key question on which our very survival may hinge. Can we continue to take from our fields and gardens the very essence of life, century after century, in exchange for a small portion of sterile mineral matter, and feel we are paying our due?

In our consumption of a plant as food, we utilize only about 25% of the fertility of the soil used in producing that food. In other words, in the route from soil to plant to animal and/or man, 75% of the fertility of the soil is retained in the part of the plant or animal that we refer to as waste. This is what we bury in our garbage dumps and flush down the sewers. Yet this is material that nature does not consider waste, but in her way of doing things, would return to the earth to replenish new life. In nature's pattern, the end of one life is the beginning of another. But our wastes are buried, burned and dumped into the rivers or destroyed.

The crisis in agriculture is not just the farmer's problem. I'm afraid we are all involved. He can only begin to pay this huge debt to nature if we all insist, and if we make it possible. It is going to demand a complete rethinking of our agricultural policy. It may involve a reversal of the population migration, a decentralization of populations, and all the resulting re-evaluation of our priorities. The problems facing the cities in the 70's may help force this reappraisal of our goals. The cities are rapidly becoming unfit for human life. We need to take a good hard look at where we are heading, and whether or not this ever-increasing gap between man and his earth is really what we want.

Such broad implications may very well leave you feeling overwhelmed and helpless as to what can be done. You are not alone in this plight. It has bothered many of us for many years. But I can tell you this: The current widespread interest in ecology is not just a passing fad. At least I hope it isn't. The science of ecology can help us because it *does* put man in his proper place in the web of life. We simply must try to understand more of God's laws of nature, act accordingly, and all things will begin to fall into place. It will require much study from all of us. As young people you are not burdened with orthodox methods and traditions. But remember that traditions do provide the framework—and often the only framework—within which our society knows how to function.

We can make the changes needed, within the existing framework, but the picture I see that needs to emerge is quite different from the one I can see today. We need to develop what we may refer to as an intermediate technology in agriculture. I'm sure we cannot and will not go back to the traditional peasant's way of farming, and all the hand labor involved. But we do need to study his methods and adopt what we can to this new technology. *The technology we develop must be governed by ecological principles.* The monoculture that results from vast areas of cash crops must once again be broken up by crop rotations. Field sizes can and will be larger and better adapted to modern machinery. But grass and legumes must once again take their place in our rotations.

Farms certainly can be larger than they have been, but they can also be small and operate efficiently if some of the larger machines are owned jointly or by cooperatives. The number of people involved in the production of our food will be greater when food requirements go up. And when one considers the number of urban people engaged in so many non-productive and service-oriented occupations, there should be no trouble in finding the people who really never wanted to leave the farms in the first place. And if the consumer wants to assure himself of an adequate supply of high-quality food in the years to come, he will have to pay for it, through either greater effort on his part or a willingness to reward those who do it for him. Somehow all of us must come to the realization that there is really nothing free. As the ecologist reminds us again and again: everything is connected to everything else. This includes the price tags.—*Robert Steffen*

CHAPTER 4

Looking For
The Organic America

Did you ever try to tell somebody something—maybe for years—and then suddenly have him come around one day and start telling you all about it? The reason I mention this is because a lot of organic farmers must be having this kind of experience nowadays. It seems like once or twice a week there'll be an article in some magazine or newspaper or a television news review examining the exploding organic food industry.

Probably for the first time in United States history, there is a market. Presently, there aren't enough all-organic farmers to even meet the demand half way. As a matter of fact, it is probable that by the time you read this, the demand will have doubled or tripled. Naturally, this condition is creating its share of confusion; some of it is genuine, some of it is not. Some farmers want to be organic farmers and some farmers just want the organic market.

This doesn't have to be as ominous as it sounds. Organic farming has its disadvantages—and with more information and a little experience, a number of would-be organic farmers will eliminate themselves. Conversely, those who want to be organic will move right into it as fast as hard work, nature, experience and information will allow them to.

What I'm finding is that a new organic farmer has to develop his operation past a certain point before his experiences will develop his confidence. Two factors seem to pop up all the time. First, he or she has got to like doing the kind of work organic farming entails. There has to be some continuing interest and pleasure in growing things, and in developing the skills and knowledge necessary to farm successfully by organic methods. Then there has to be a commitment; somewhere, at some time, some kind of decision has to be made not to go back.

Looking around and talking to part-organic farmers, I do not find that they have made this sort of commitment yet. Some are on the verge, and some may never be able to make it for themselves. As a matter of fact, backing away and taking a big look at everything I've seen so far, the picture is that the chemical-mechanization revolution has made this a nation of agricultural cripples. Important skills have been lost, substituted by specialized equipment designed to work with or harvest food also designed to be harvested by the special equipment, or to fit neatly into a can. Chemicals have more or less replaced farmers' knowledge, for farmers rely more upon what compound can be purchased than what they understand and can do for themselves. Generally speaking, we've lost our confidence.

Not long ago I was visiting with a young man whose youth, education and opportunity should have driven him head-on into a magnificent challenge. Sorry to have to report that he saw no challenge. He is subjected, and he sees his subjection and accepts it bitterly. In the bloom of his youth he is sour, dissatisfied, disgruntled, whipped by a single species of insects—that is, whipped by the lulling, secure, easy, crippling knowledge that parathion will kill.

Now here's the thing that leaves you kind of disturbed. This young man's family came to this country as immigrants, somehow established themselves on their own land, and successfully farmed without chemicals for many years; yet, he said, speaking of parathion, "It's the only way we can control these insects," and that is the peculiar dichotomy I'm seeing everywhere. He knows that his parents and grandparents farmed successfully without using destructive chemicals; he knows that such chemicals are damaging our environment and destroying a balanced ecology; he believes that eventually we'll have to find a way to get along without them; yet, he believes in parathion. Why is he so bitter? Because he's waiting for someone else to solve his problems.

* * *

The contract between Chico-San, Inc., and Wehah Farm is probably a landmark which spells out the way the industry will take to prosper and develop.

Chico-San is a distributor of macrobiotic foods. Brown rice is their most important product. The company was established in the prime rice-growing region of Chico, California, because its president, Robert Kennedy, felt that the finest organic rice would be produced in that area. But it took eight years for Kennedy to locate his producer.

The four Lundberg brothers, who own Wehah Farm, were interested the first time Kennedy talked to them. Interested but concerned; a number of things had to be weighed which also involved their neighbors and a life-long way of handling their business. It was not an easy decision.

Wehah Farm is located in Richvale, California, which is a small, compact farm community 20 minutes' driving time from Chico. The rice farmers there are lifelong friends and neighbors, and the farming nucleus of the community centers around a co-op association which mills and distributes the rice grown in the area. Apparently the members are passionately loyal to the association and view anyone contemplating withdrawal as contemplating a personal rejection of all members.

There were other serious matters to be considered. Leaving the association meant leaving an easy, almost automatic market. Leaving chemical rice farming meant leaving an easy, almost automatic farming method. Leaving meant leaving a substantial bird in the hand for two very dubious birds in the bush. Modern rice farming is a costly operation. The Lundberg brothers were not certain that they could raise enough rice organically to pay the way. But Kennedy had something working for him, something instilled in the hearts and minds of the Lundbergs.

Mr. Lundberg, the father of the brothers, had been an organic farmer before his time; that is, he held to certain ideals which form the basic principles of organic farming. He taught his sons that a farmer has an obligation to improve the land which he uses, and if possible, a farmer must leave his land in better condition than when he took it over. He also impressed upon his sons the value of cooperation, of cooperating together, and together cooperating with their community and with other communities. Wehah is a symbol of family cooperation—Wendell, Eldon, Harlan, and Homer.

Every time Kennedy talked to the Lundberg brothers they listened with their hearts and tried to make it work out in their minds. Modern rice farming is so sophisticated and chemical that the practicalities of considering organic farming were frightening. It's a long, long way to make such a change.

And all the while buried in their hearts and minds was their father: "Improve the soil. Leave it better than the way you found it."

They decided to try. And 76 acres were set aside and prepared organically. A source of manure was located, the manure composted and worked into the soil. An organic rice crop was harvested which averaged 37 hundred-pound bags per acre, which is quite low by chemical standards, but high enough, with organic prices, to demonstrate economic feasibility. The brothers made up their minds.

9

They discarded the chemicals, withdrew from the co-op association, rejected their agricultural educations, their years of experience, and committed their 3000-acre Wehah Farm to organics.

Prevention through immaculate housekeeping provides the major control system. Tadpole shrimp are controlled by careful water manipulation and timing. For example, if the rice is seeded by spreading the seeds upon the dry ground with a truck and then flooding the paddies, the rice has an even chance, at least a week. Tadpole shrimp and rice seedlings are developing at the same time. Raising and lowering the water level helps to flush away silt and reduces, for a time, the distance which the chlorophyll-producing sunlight has to travel through water.

Seeds are not treated to prevent seedling diseases. The Lundbergs' preventive method is to use healthy seeds. Rice water weevil infestation normally occurs within 10 to 20 feet adjacent to the levees. The controlling preventive method used by Wehah Farm is clean and careful housekeeping, disking and cleaning away weeds up to and around the banks. Rice leafminer does not seriously harm healthy, vigorously growing plants, especially when most of the leaves are off the water. Fertile soil and water manipulation solve most leafminer problems.

Mosquitoes are effectively controlled biologically through a special arrangement with the Butte County Mosquito Abatement District. The flooded rice paddies are planted with a small insect control fish called gambusia, which feeds upon the mosquito larvae and numerous other unwanted insects as well. The fish thrive in the Wehah paddies and are a living verification of their organic methods. As a matter of fact, the brothers do not worry too much about insects now. Weeds are their next big pest problem to solve.

They expect to work out most of their weed problems by utilizing good organic farming methods. Crops are rotated. One year rice, the next year barley or some similar grain, and the third year a paddy is allowed to lie fallow. Eventually, the fallow paddy is irrigated to start the watergrass and other weeds, then it's drained, the land disked, and a cover crop of purple vetch is planted to protect the soil and supply a rich, natural source of nitrogen. In the following spring, composted manure, especially chicken manure, will be worked into the soil prior to planting.

Everything goes back into the ground except the grains of rice. To someone who has never seen rice stubble and straw this might be difficult to appreciate. Rice straw and stubble are extremely tough. Rice farmers burn their harvested fields because it is expensive to till the soil properly and work the straw and stubble in. A conventional disk will likely roll over more straw and stubble than it will cut through and turn. Eldon Lundberg relates how difficult it used to be. "We just had to get out there and fight it." Nowadays, it's a snap. First, paddies are irrigated right after harvest to start the decaying processes as soon as possible. In the spring, the ground is worked over with a giant-sized stubble disk (tandum disk). This is a no-nonsense piece of equipment capable of chewing up a two-by-four. To give an idea of its size and heft, it takes a 320-horsepower D9 construction cat to pull it through the paddies.

Organic rice farming is expensive. Harlan Lundberg estimates that their operational cost runs almost twice as much as their neighbors, which only tells part of the story. To date, Wehah Farm produces an average yield of 37 one-hundred-pound bags per acre; chemical farmers are getting 70, 80, and sometimes a hundred bags per acre, which means that presently Wehah Farm pays out twice as much to get half as much. The yields will improve as the soil is improved. As a matter of fact, they're hoping to hit about 50 bags per acre from this year's harvest. Naturally they'd like to get the yield up, not only because they'd like to increase their profit and offset their high operational cost, but because they would like to get the price of organic rice down.

*　　*　　*

Frank Ford is a dry-land farmer. A portion of his 1,800-acre farm in the Texas

10

Panhandle has enough ground-water for irrigation, but he feels it would be too hard on his soil—and if this gives you an idea of what kind of farmer he is, you're right. He's the best kind. "If you fight nature in farming, you're going to lose; it might take 20 years to lose, but you're going to lose. Whereas if you work with it, then every year your soil is stronger, your plants are stronger, finances are stronger as a result."

On all accounts Ford can demonstrate his point. When he first took over, much of his land was badly eroded; "There were gullies so deep that you could hide a tractor in them." The farm was badly infested with Johnson grass. Nowadays the gullies are gone; they've been filled and the land terraced and leveled so smooth that an amateur can operate his tractors. The Johnson grass is gone and very little precious water runs off from his soil.

Frank would laugh at the idea of hiring a soil management system. Summing up what he said about farming boils down to this: "A farmer is first a soil manager or he is no farmer." A Texas A&M graduate, he relies more upon practical experience and a deep commitment to organics than upon flashy fertilizers or hypothetical methods. Eleven years ago he stopped using pesticides, "and I have had many infestations of green bugs and brown mites . . . ladybugs always do the job . . . I would hate to kill my ladybugs in a moment of panic."

Non-organic farmers use 2,4-D or an assortment of herbicides to control weeds. Frank and the farmers I talked to use skill. Johnson grass, for example, is very easy to get rid of with the right equipment and good timing. Many of the chemical people, of course, have been around trying to scare the farmers into treating their seed against wireworms and rust. According to Frank, it's a myth, "I have always used seeds that I would have eaten—and have many times."

Around West Texas, dry-farming 1,000 acres is about a good hefty one-man farm (1,800 acres is more like a two-man job) except for harvest time, which would make it seem like enough work for anybody. But years ago Frank acquired the idea that he'd like to have a hand in merchandising the fine hard TASCOSA and STURDY wheat that they grow down around Hereford, so when he had finished his tour of duty in the service, he contacted a small mill owner in town and the two got together. You might say that he more or less started his farming and milling careers simultaneously.

In 1960, Ford reorganized the mill with George Warner, an agronomist who was also putting together a seed company, and a few other farmers, incorporating the company as Arrowhead Mills.

Using the formula of quality products, hard work, and fair prices, Arrowhead Mills has prospered and gathered a following among growers—or perhaps it would be better to say that a number of people who share similar ideas have come together and organized the Deaf Smith County Organic Farmers Group. The mill provides the practical impetus by paying a published premium for organic grains. However, the focus of the group centers upon organic farming methods, the vital need to protect and to improve the soil of West Texas, and a very conscientious desire to grow better food.

Working with the group, as well as operating his composting business, is Fletcher Sims of Canyon, Texas. As a former student of Dr. William Albrecht, Fletcher uses Pfeiffer's Biodynamic method of composting. In these *still* early days of volume composting, Fletcher Sims is a pioneer who has had to design his own techniques and much of his equipment. He has an interesting goal—or is it a mission? He estimates that the feedlots in the county produce enough manure to put a thousand pounds per year on every irrigated acre in the county, and that's his goal.

Organic farmers in the county who sell to Arrowhead Mills agree to a soil-conditioning program which incorporates Fletcher Sims' compost and feedlot manure. The first year, for example, they use a thousand pounds of compost and six tons of feedlot manure per acre. The compost, rich in microorganisms, is applied over the manure just before the field is harrowed or swept. The microorganisms in the compost rapidly decompose the manure and nitrogen tie-up is held to a minimum.

Meanwhile, way up on the northern edge of the Panhandle in Gruver, Don Hart is composting and he'll supply farmers in that area. Don operates a small feedlot and farms 2,000 acres of irrigated land. About four or five years ago Don began to worry about the abrasive conditions brought about by irrigation and commercial fertilizers. The land was tightening up; yields were down in spite of irrigation and hefty doses of synthesized nitrogen and superphosphates; and the more he looked around, the more he worried. These days Don will tell anybody flat-out that farmers in West Texas have two alternatives, either learn to manage their soil properly or create large areas of wasteland.

When he started worrying hard he began looking for solutions—which has led him straight into organics and composting. Two years or so back he heard about Fletcher Sims, who eventually got him started composting biodynamically; now the two work closely together. Fletcher in the south counties and Don in the north. It was inevitable that Frank Ford and Don Hart would also get together, and it's very likely that Don will be forming an organic farmers group in his county. He has a lot that he can say—and something to show for what he's talking about. Just driving alongside one of his fields and comparing it with his neighbor's will stop most arguments.

I saw fields of chemically fertilized irrigated grains in West Texas that wouldn't make a respectable dry-land crop in states from North Dakota to California. And when Don Hart and Fletcher Sims talk about how farmers down there are getting scared I've got to wonder what's been taking them so long.

*   *   *

It's news when a major food firm takes action to bring a completely natural product to the American market. It's even better news when that company has a long reputation for absolutely top quality in its field. Put both those factors together and you'll realize why the Sacramento Foods story is a significant one.

A division of the vast Borden Foods complex, Sacramento has initiated a canned breakthrough: *It is ready to market organically-grown tomato juice and whole tomatoes.*

The cannery was established by Tom Richards Sr. in 1931 and was owned and operated as a family business until 1968, when Richards retired and the firm was sold to Borden. Borden very wisely established Tom Richards Jr. as the new president of its new division and the family style of operation has continued.

The first crop of organically-grown tomatoes was nurtured by Jack Anderson of Anderson Farms. He's representative of a lot of farmers in California these days who have been sniffing the wind and figure that it might be a good idea if they start learning something about farming organically. Basically Jack is a conservative farmer; by practice and reputation he has used commercial fertilizers and pesticides sparingly—not as a matter of commitment but as a matter of hard business. He's simply not going to be panicked into clobbering his fields with expensive chemicals. "When somebody tells me that I need something, I want to see first that I do."

The main thing worrying growers in the area is worms. By law, canning tomatoes cannot exceed a two percent worm damage. Most canneries set the tolerance down, even as low as one-half per cent. While nearly all growers think that the required tolerance cannot be maintained without pesticides, Jack Anderson was convinced that with or without pesticides, timing was more the factor. "In this section experience has taught me that if you get tomatoes out before September, there's no worm problem; after that, there's not much you can do. I've sprayed three times in a September and still had worms."

This particular field had been allowed to go fallow last year, so there was a nice cover crop to plow in; the year before that it was in wheat, and before that in safflower. The field had been hand-hoed twice, and weeds growing up between the vines were topped (cut) before their seed could mature.

The first thing to do when walking into a field of tomatoes which is supposed to be organically-grown is to stop and listen. If it's around the month of July and no pesticides have been used, the field hums, kind of a buzzy-buzzy singing. If it has been sprayed, it's dead silent. (If you want an eerie feeling sometime, walk through an orchard that's been sprayed.) To leave no doubt, if insects cannot be heard, some should be easy enough to find with a little looking around and shaking of vines; otherwise the field has been sprayed or dusted.

And if a field is high in natural nitrogen, or if it's been dosed with synthesized nitrogen, tomato vines will be a rich bright green. However if the nitrogen content is just adequate, the vines tend to be a little dusty looking, a grayish-green without the bright hue.

Anderson's field passed on all appearances. Residue tolerances were satisfactory. The contract between Anderson Farms and Sacramento Foods stipulates that if commercial fertilizers, herbicides or pesticides are used, the agreement would be null and void.

After I left the cannery I went up to Woodland to visit an organic farmer in the area. He knew all about the tomatoes and Jack Anderson. "That field has been watched closely. Jack wouldn't fool around; if he says he's going to do it organically, that's what he'll do." The tomatoes were organically-grown.

Sacramento Foods has set a basic policy that opens up an opportunity for other firms to follow. Here's one major food processor that's not letting today's most urgent opportunity pass it by—and without question has taken a significant lead.—*Floyd Allen*

PART II

ORGANIC FARMING METHODS

CHAPTER 5

An Interview with a Nebraska Farm Manager

Bob Steffen graduated from Creighton University in Omaha. He's tried to keep his methods efficient and economical on the 1,000-acre farm he has managed organically for Boys Town for almost 25 years.

Q. What kind of special equipment do you have on an organic farm?

A. No different from what any other farmer has. We may use a manure spreader more than most . . . and we don't use the spray rig or the fertilizer drill for putting nitrates in when planting. There's no need for special equipment. We use a chisel plow and sweeps instead of a moldboard plow.

Q. What goes into making fertile soil?

A. When it comes to explaining soil fertility and plant nutrition, I always think of an article Dr. Ehrenfried Pfeiffer wrote back in 1957 in *Natural Food and Farming* magazine. He explained how an excess of one fertilizer element may cause deficiency symptoms with regard to others, even though these others are theoretically present in sufficient amounts. Excess symptoms in soils and plant nutrition may not be recognized as such, but treated as deficiencies of something else.

> "The example of excessive nitrogen fertilization is at present probably the best known. Plants take up too much nitrogen and begin to show deficiency symptoms of phosphate and potash. The fertilizer-minded farmer then is advised to use more phosphate and potash, while the restoration of the balance with less nitrogen would bring about the same effect—better production.
> "This nitrogen situation, however, involves another problem: that of the protein quality versus protein quantity. It is a general occurrence that the protein production in a plant is reduced with the depletion of the soil. Many figures in nutritional textbooks with regard to protein content no longer apply. For instance, the protein content of wheat (red, soft) is given as 12%. In fact, most red wheats analyze today between 8% and 11% protein. In corn, it has been found that more nitrogen increases the protein content, but it has also been discovered that the protein quality is lacking."

The idea of "balance" in speaking of soil fertility cannot be emphasized too much. A balanced soil presupposes a living, viable soil which is more or less stable and has adequate reserves, with high humus content. The problems of imbalances that ecologists are so upset about today were clear in Dr. Pfeiffer's mind 30 years ago. The emphasis on quality in an age that is so engrossed in producing quantity is indeed refreshing and long overdue.

Incidently, Dr. Pfeiffer's reference to the protein content of grains is even more relevant today. We know of corn samples from high-yielding fields that tested as low as 5 per cent protein. We had a corn sample from one of our fields in 1966 that tested over 14 per cent protein.

Q. How many tons of organic material do you put on your land?

A. We will put sometimes as much as 20 tons on an acre, but the average is about 10 tons of compost-treated manure. Fall is the best time for spreading.

Q. How do you prepare yourself psychologically to not using sprays?

A. Let me try to get specific on that one. At Boys Town, we have a corn root worm problem, but is has never got to be an economic problem. Now most farmers feel if they don't use some chemical to control it, they'll be in deep trouble. Chemicals give them the idea that they're doing something. Actually once they get the idea that they're doing something *better* by building up their ground organically, good things will begin to happen for their income and our environment. We feel at Boys Town that we're better off in the long run by not using chemicals for such a problem as corn root worm. Now we hope we have convinced other farmers to give these methods a try.

Q. What about soil tests?

A. When we first started using the organic method 25 years ago, we tested every field and did it quite regularly. We would make soil tests on the fields that were going into alfalfa, which is one of our main crops. It's expensive to sow alfalfa. It's a small seed and everything must be just right. The soil test is important to establish the seed. But once built up, soil samples need only be checked occasionally. Organic content should be well over 1.5%; when below 1.5% problems will appear.

Q. Where can you find enough organic material?

A. Farmers can find sources of organic matter by the tons wherever they live. Everything from municipal wastes like treated sewage sludge to cover crops. If you use your imagination, and make a few phone calls, you'd be amazed. Around Omaha, we get manure from the stockyards, treated paunch wastes, etc.; elsewhere, they might be local canneries willing to take huge amounts of valuable humus to your farm. Solid wastes become valuable once you handle them properly.

Q. What about putting crop residues back on the land?

A. *Crops and Soils* magazine recently had a fine article on this subject by L. S. Robertson and R. L. Cook of Michigan State University. They pointed out how soil organic matter has been known for years to be "the life of the soil." Thirty years ago, Ohio scientists showed that over a period of 33 years of cropping, relative corn yields decreased in proportion to the decrease in the soil's relative nitrogen content. They used nitrogen percentage as a measure of the total soil organic matter. Other scientists showed similar relationships.

Despite these significant studies, many agronomists and livestock husbandrymen still echo such words as: "Don't waste your crop residues. Put them into the silo, feed them." Actually these crop residues should be incorporated with the soil to feed the countless organisms living there.

"High levels of active soil organic matter are essential for high yields," say soil scientists Robertson and Cook. ". . . Perhaps the easiest way to add organic materials to the soil is through winter cover crops."

CHAPTER 6

From Where Comes Soil Fertility?

THE UNITED STATES is blessed by having within its borders large areas which produce food in great abundance. Several of these stand out because of the high quality and high nutrient content of the food which is grown therein. These latter areas include the corn belt across Illinois and Iowa, the high-protein wheat belt in Kansas, the grain and fruit areas in Oregon and Washington, the fertile "Great Valley" of California, the major "river bottoms" or the flood plains of the large streams which flow across our country, and the stock grazing areas of our western states. Other areas might also be included.

What do these areas have in common which may be responsible for the nutrients in their soils? Can it be climate? The corn belt is in a region of moderate to heavy rainfall, with hot summers and cold winters. The surface of the land is flat to rolling. The elevation is low, being only a few hundred feet above sea level.

The Oregon grain land, however, is in a radically different climatic setting than the corn belt. In the dry farm country, a crop is recovered only every other year. Each field of grain this year is matched by a counterpart which lies fallow and picks up its scanty measure of moisture which is held for the next year. The growing season is entirely different from that in Illinois.

The Washington fruit country is still different, and the "Valley" of California is unlike all the rest. Irrigation is necessary here. Even in California alone the temperature ranges through a cool wintry season in the north to a sub-tropical climate in the Imperial Valley to the south.

Clearly then, temperature and rainfall are not the common denominators in the development of fertile soil.

If climate is not the common factor, perhaps their soils have a common heritage in abundant black humus as occurs in the dark, corn belt soils. Or, are all these soils at the same stage of maturity, or are they texturally favorable so that they will "raise anything?" The answer is definitely "No" to all of these questions.

Whereas the corn belt soil is typically rich with dark humus, the Oregon soil is grayer, and some irrigated, desert California soil may be so light colored as to appear as if it contained no organic matter. The soils range from a mature stage of development to others where a profile is hardly apparent. They vary in texture from heavy clay to almost gravel. Their common denominator still has not been found.

The solid rock beneath the Oregon-Washington rich soils is commonly basaltic lava, but the deeply buried, solid country rock in the corn belt region is sedimentary rock, such as limestone, sandstone or shale. In the grazing areas, the country rock may be sedimentary, or it may be igneous granite. Under the alluvial stream bottoms it may be any kind. But here we come to a clue. The fertile soil is not a product of the solid, underlying rock; it may develop from a stream-washed or other secondary deposit above the solid rock. Let us look into that possibility.

The parent material underlying the soil of the corn belt is a glacial deposit

19

which is more or less covered by a wind-blown mantle. The well-known, rich Palouse soil of the Oregon-Washington region is derived from wind-blown silt, clay, and sand which overlies the solid rock. In the "Valleys" of California, thick alluvial deposits built of rock waste which was carried in from the surrounding high mountains rise high in the intermontane troughs, and these are the parental sources of the soil. The alluvial flood plain deposits in the river bottoms give rise to their fabulously rich soils. The thinner but nutrient rich soils in the western grazing areas may lie on rocky, mountainous slopes, without benefit of depositional processes which accounted for the other thick deposits.

As far as superficial classification goes, these parental sources are different: glacial, windblown, alluvial, and slope washed. Is there anything in common? The answer, if we look closer at the constituents of the fertile soils is "Yes, all of them contain *finely pulverized, little weathered, native rock and mineral particles.*"

All of the before-mentioned areas of fertile soils are developed upon parental geological deposits which contain abundant physically weathered (little chemical weathering), finely pulverized, well-mixed native rocks and minerals that contain a reserve of nutrient elements in an adequately available condition. We will review them, one by one.

The corn belt soils are underlined by a thick glacial deposit which was brought in from the north and from nearby rocks, by a thick, grinding, churning glacial ice sheet that later melted away and left its crushed and pulverized rock load fabulously rich in slightly weathered rock and mineral particles. Let us contrast the situation here with that of an old leached and worn soil. In the latter soil, usually only barren quartz (silica), sand and silt, coated with a thin film of reddish to brownish iron oxide is mixed with some exhausted, degraded clay, and leached gravel fragments composed of chert or flint. Nothing of nutrient value to plants remains.

Not so with the glacial deposit of the corn belt. The ice sheet which moved southward from Canada, across the Great Lakes Region, picked up, shoved, carried and mixed mechanically the igneous (heat-formed) rocks of the north with the sedimentary rocks to the south, and sweetened them with various concentrations of copper, zinc, cobalt, manganese, iron, and other ores (trace elements) over which it occasionally passed.

The resistant potash feldspars of the igneous and metamorphic rocks were mixed with potassium-containing sedimentary shale which releases potassium easily. Calcium and sodium igneous feldspars were mixed with the calcium-rich, easily soluble calcite of sedimentary limestone. Magnesium, calcium, and iron in crisp, hard, hornblende and pyroxene minerals were blended with magnesium, calcium and iron in sedimentary, easily soluble dolomite. Slowly soluble phosphate of calcium in the igneous mineral apatite was churned alongside of quickly soluble phosphatic sedimentary limestone and shale.

All of these rocks and minerals were crushed by the grinding and abrasive work of the ponderous, slow-moving, but relentless heavy ice sheet. They were pulverized to particles so fine that the melt waters from the glacial ice were milky white with rock flour, yet these microscopically fine particles were unchanged, unleached, and as rich as the original rock in mineral nutrients! This point is most important; the unchanged, native rock and mineral particles were pulverized under icy, cold, non-reactive chemical conditions and were thoroughly mixed and blended with all other kinds of rocks and minerals, so that the final product constituted a rich potential feast of a *wide variety of inorganic nutrients* to be *chosen as desired* by *plant rootlets*, in a wide range of availabilities. The fine pulverization of the particles made them easily susceptible to attack by plant rootlets, and the particles had not lost any of their nutrient constituent elements because they had been pulverized mechanically in the cold, without undergoing chemical weathering or loss.

It is no wonder that the corn belt soils have high reserve fertility. Think of the

20

fertility reserve, silt-size minerals which can be drawn upon year after year and crop after crop. The pulverized glacial product is a naturally perfect, fully blended agstone, widely spread for development into soil. *The modern farmer who wishes to add a balanced agstone to his depleted soil need only duplicate in his crushing plant the formula that Mother Nature worked out and produced from her glacial mill! There are no patents on this fertilizer.*

The quick development of a fertile soil on the glacial deposits included a favorable climate for organic growth, of weathering to a high-exchange type of clay mineral, and flat topography which did not lose by erosion the fertility that developed.

The Palouse soil parent of the northwest states was redeposited by the wind after a glacial ice sheet pulverized the surface of the widespread basalt rock flows which underlay the region. Again, the native, unweathered, nutrient-rich basalt was mechanically pulverized by the ice, then picked up, mixed, and scattered far and wide by the wind, and deposited to give rise to a fertile soil today. Finely pulverized, native rocks and minerals are the crux of the fertility.

The Great Valley and the lesser intermontane valleys of California owe the inorganic fertility of their soils to mechanically pulverized and weathered rock fragments that have been washed in by water, blown in by wind or carried part way by glacial ice to their present resting place. The arid climate has preserved without chemical weathering or leaching the nutrients in the fine rock particles. Upon being irrigated, the moisture, and the weathering effect of plant rootlets now releases and pulls out the stored-up, adequately available potassium, calcium and other nutrient elements. To repeat, the native rocks were reduced in size, the arid climate protected and preserved the pulverized particles, and man is now capturing the inorganic fertility prize which long awaited development.

The soils of eastern Colorado and western Kansas, notable for high protein food production, have developed upon deposits rich in unweathered native rocks and minerals which were derived from the Rocky Mountains, and were washed out eastward on to vast, wide-spread, coalescing alluvial fans. These alluvial deposits thin eastward, downslope across western Kansas. Tremendous quantities of rock fragments were washed away from high, primitive Rocky Mountains. The unweathered to partly weathered rock fragments scattered across the plains bear mute testimony to the process of deposition just described.

Fertile soils in river bottoms, and the wind blown loess which occurs on adjacent valley walls are replete in silt and clay size particles of unweathered to partly weathered native rock and mineral fragments. These fragments were washed down by the rivers, or were pulverized by the abrading stream, and then spread out over their flood plains (bottoms) in time of flood, or on their sand bars during high water. Wind blowing across the dried deposits picks up the silt and clay, carrying them upward to the land alongside the river. The theme here again emphasizes unweathered to partly weathered native rock and mineral particles, with clay and organic matter, carried by a geologic agent and deposited where soil can develop upon it.

Thin, but nutrient-rich soils in the stock grazing ranges of our western states develop upon the primitive underlying rock, or they accumulate from slope wash and gravity creep of the mechanically weathered rock in the higher elevations. Here, as before, the nutrient elements of the native rocks and minerals are the sources of the inorganic fertility of the soil.

Numerous examples from Nature show how soil fertility has been evolved over the past. Man may duplicate artificially, on any size scale which is consistent with economics, the same process of developing fertility which has been so manifestly successful before.—*Dr. W. D. Keller*, Dept. of Geology, University of Missouri.

Glossary of Geologic Terms:

Alluvial—Relating to mud, clay, or other material left by running water.

Alluvial Fan—The alluvial deposit of a stream where it issues from a gorge upon an open plain.

Abrade—To scrape, wear, or rub away or to remove by friction.

Calcite—A calcium carbonate occurring in many crystalline forms, such as chalk or marble.

Dolomite—A calcium magnesium carbonate of varying proportions found in limestone and marble.

Feldspar—A closely related group of minerals, all silicates of aluminum, with either potassium, sodium or calcium.

Hornblende—A common dark mineral, generally black, composed of a light to dark green silicate of iron, magnesium and calcium with aluminum.

Igneous—"Fire formed" rock, the result of the cooling of molten rock below the earth's crust.

Loess—A fine grained, erosional sediment, deposited by wind, covering vast areas in Asia, Europe and North and South America.

Iron Oxide—The combining of iron with oxygen—rust.

Pyroxene—Any of a common and important group of iron, magnesium, and calcium silicates found in many igneous rocks and molten lava.

Sedimentary—Designating rocks formed by simple precipitation from solution, as rock salt and rocks formed of organic materials, as limestones, coal and peat.

Weathered—Rocks whose appearance is changed by exposure to the atmosphere.

CHAPTER 7

Using Rock Fertilizers on the Farm

ROCK FERTILIZERS AREN'T GENERALLY hard to ship, handle, store or spread. They can be handled the same way as other bulk fertilizers for farm application. Although they may not always flow like pelleted material, the same spreading machines and other equipment will usually distribute all of them effectively. One important point is that they are not corrosive on metals.

Rock phosphate should always be bought from a source with as low a fluorine content as you can get. Fluorides may become a problem after many years of using any material high in this toxic element.

Generally speaking, the finer the grind of rock phosphate, the more available it becomes to the growing plant. This does make the material rather dusty and hard to spread on windy days. If you can get a bulk spreader truck with a hood covering the spinners, it will enable you to do a better job of covering your field, especially on blustery days.

If you receive a bulk carload of rock, it can be stockpiled and spread later. Rain will not cause it to cake. It can be spread with the trailer-type spreaders that you might rent. The rock can be spread any time of the year—with autumn usually the best time to incorporate ground-rock fertilizers. The work load is less, the fields are all open, and the heavy traffic on the fields generally causes less compaction because the soil is usually drier.

Furthermore, the winter freeze will undo some of the damage from compaction. Rock fertilizers can even be spread on snow, as long as the ground is not frozen and the machines can move where they need to. Fall is similarly the best time of the year to spread compost or manure, for pretty much the same reasons.

This preferred timing is really a strong argument in favor of composting—because the manure is stockpiled through the winter and wet spring when one cannot easily get on the fields, and through the summer while growing crops are on them. This will give you the compost you need when you should spread it, and in the quantities that you should have.

The discussion above applies pretty much to potash rock as well, or to any other natural sources of mineral fertilizer. Potash rock, however, is generally not as fine as the phosphate and may actually be easier to handle in some respects. What we have seen and used at the farms at Boys Town has almost the same characteristics as limestone.

Ground-rock material can, of course, be shipped in closed, hoppered cars in bulk, and it is available in sacks, which would enable you to spread it with any grain drill-type spreader. Both the cost and the labor involved would be greater. However, in some cases, this may be the only way you can get it on your land.

One other point that may be brought up here is that if soil tests, tissue tests, and your observations indicate trace mineral deficiencies in your area, this can be a good way of overcoming them. Simply mix your own formula with some rock fertilizer and spread with the grain drill-type spreader. This is more accurate, and the amount of material involved is usually only about 100 pounds per acre of the complete mix, enabling you to put on the very small amounts, and put them where you want them. Generally speaking, this may involve only a pound or two of some material per acre. Mixing these with the more bulky rock is a good way to get your fields covered.—*Robert Steffen*

CHAPTER 8

Legume Nitrogen

LEGUMES IN ROTATION with proper inoculation are a very inexpensive source of nitrogen for any organic farmer.

All that is needed is a legume suited to your particular climate and soil, and bugs. Not ladybugs or wasps, but microorganisms like *Rhizobium trifolii*, which are available at seed stores in lots of three to four billion. This combination of legume seed and bacteria inoculation at planting time will reward you by adding up to 400 pounds of nitrogen per acre per year to your soil, organically. And when you are ready to change crops you utilize this added nitrogen and receive another bonus from the plant residues. Green manure from leguminous crops contains much higher amounts of nitrogen than from nonlegumes. Sweetclover at its peak contains over 3% nitrogen compared to less than ½ of 1% nitrogen in wheat straw.

The amount of nitrogen and organic matter produced through legume inoculation will vary between locations and crop species. The table gives the average amounts of nitrogen fixed per acre per year by some important legumes.

	lbs N/acre	*lbs/acre*
Alfalfa	194	15-20
Ladino clover	179	1-4
Lupines	151	50-120*
Sweetclover	119	10-15
Alsike clover	119	6-8
Red clover	114	8-12
Kudzu	107	10-15
Legumes in pasture	106	†
White Clover	103	1-4
Lentils	103	12-15
Crimson clover	94	15-20
Cowpeas	90	20-30
Lespedezas	85	10-15
Vetch	80	35-80*
Burclover	78	20-25
Peas	72	50-60
Velvetbeans	67	30-40
Soybeans	58	45-60
Winter peas	50	50-80*
Peanuts	42	40
Beans	40	25-80*

*Depends on species and method of seeding
†Rates vary in mixtures

(1) Alfalfa—Alfalfa, yellow alfalfa, sweetclover, black medic, burclovers, and button clover

(2) Clover—Red, white, crimson, alsike, strawberry, sub-terranean, hop

(3) Peas and vetch—Peas (field, garden, Austrian winter, sweet), vetch (common, hairy, purple, narrow leaf), lentils and horse beans

(4) Beans—Garden, scarlet runner, kidney, navy, pinto, great northern

(5) Soybeans—All

(6) Cowpeas—Cowpea, partridge pea, pidgeon pea, lespedezas (sericea, Korean, common), crotalaria, kudzu, peanuts, indigo, velvet bean, lima, and Mung bean

(7) Lupines—Blue, yellow, white

(8) Specific groups—(a) Bird's-foot trefoil, (b) Big trefoil, (c) Crownvetch, (d) Black locust

Some people feel that a pound or two of white clover is not a very hefty application and want to put on more. Don't. All you will do is overpopulate and weaken the stand. White clover seed is very small, and there are 800,000 seeds per pound (compared to 200,000 alfalfa seeds per pound or 1,000 peanuts) and this is what determines the different seeding rates. Most small-seeded legume-recommendations are twice as high as would be needed if all the seed survived. Mortality rates are usually very high, and if you do a good job of seedbed preparation and seeding and have favorable weather you can get by with much less seed.

Always study the tag when you buy seed. In many states the laws governing seed sales seem to have been written by the seedsmen rather than the farmers. For instance, there may be no limit to the amount of weed seed, trash, hard seed or there may be no minimum germination guarantee. Don't try to bargain hunt when buying seed unless you are reading the tag. For instance, seed selling for $20 per bushel having a germination of 60% would be a bad buy compared to seed selling for $30 with a 92% germination. The $30 seed would give you more live seed per dollar.

Once the proper legume for your location has been selected, you must also select the proper inoculum. Not all rhizobia are alike. Some types are not as effective as others, and often low yields can be attributed to the use of the wrong type. The following plants are listed in groups that may be inoculated by the same type of microorganisms.

Success in growing legumes depends on other factors. Legumes tend to be more sensitive to soil reaction than other crops, and a neutral to slightly acid soil reaction will usually favor legume growth. Legumes are also more sensitive to the supply of phosphate, potassium and calcium than other crops. They are soil-improving crops, because in addition to supplying their own nitrogen, they apparently are not as efficient at extracting some other nutrients and therefore do not deplete the land. And, of course, they improve tilth.

Small-seeded legumes are often sown in the fall with good success; however spring seeding is generally favored. Spring is the natural time for plants to sprout and grow and when sown in the spring the tender young seedlings do not have to survive the freezing winters in the cold regions. Freezing and thawing is especially hard on top-rooted seedlings. March is a good month for seeding in the corn belt region but excellent results may be obtained earlier or later. Mid-summer is a poor time to sow legumes.

The time of sowing will determine the amount of seedbed preparation you are likely to do. Seeds are often sown into small grain crops in grass sod, on snow or frozen ground and good stands obtained. However, it has been shown experimentally that the greater the percentage of seeds that are covered, the better the survival rate will be. On the other hand, covering too deeply is often worse than no coverage at all. Small seeds, especially, are often unable to penetrate the surface when placed only 1½ inch deep; ½ inch is a good average depth for small seed.

When sowing on the soil surface or in small grain, a hand-carried crank seeder is a satisfactory method of broadcasting seed. This method may also be used on prepared ground (plowed, disked) and results will then be improved if a cultipacker is used after seeding. One of the best means of securing good stands of forage legumes in prepared ground is by means of a seeder mounted directly on a cultipacker. A drill may be used with excellent results on large seed (soybeans or peas, for example) or small seed if you are careful not to cover the seed too deeply. A disadvantage to drills is that the seed box is usually in front, and the small seed are dropped in front of the disks or shovels, and covered too deep. Study your drill, and if this is the case, try to alternate it to drop the seed behind the disks where they can be lightly covered by the drag chains. Special grassland drills are also available, but are usually expensive and give variable results, depending on the conditions they were developed for.

What are the secrets of good legume stands? I think they are the same as for all other aspects of good farm management. Timing and skill are most important and it's hard to separate the two. You have probably noticed the farms where everything seems to work out just on time. The ground is plowed one day in spring. Within two weeks the pale green shade of the corn can be seen in the morning and evening. Come fall, the corn is picked and often the drill is sowing the grain in the same field behind the corn picker. This type of farmer never leaves the ground uncovered when he can help it. He'll sow the legume in fall or spring, right in the wheat, the grass in the early spring. When the wheat is combined, the field once again will turn green as the legume grass thrusts through the stubble. Everything is done on time and with great skill.

This skill must be learned. To obtain good stands of legumes, we could list some pitfalls that the good farmer somehow manages to avoid.

1) Don't sow at the wrong time. The wrong time varies from location to location, but it is always the time when seed will not grow well. Avoid the freezes, the dry of summer and look for sufficient moisture for young seedlings.

2) Don't cover too deeply on prepared ground, and on unprepared ground use an excess of seed to allow for less favorable conditions.

3) Always select good seed, not on price alone, but on the basis of cleanness and germination. Always inoculate.

4) Maintain the soil fertility, particularly rock phosphate and potassium, and remember that pH must be neutral to slightly acid for legumes. —*James Foote*

CHAPTER 9

Crop Rotations

THIS ROTATION IS BASED ON the grains and needs of the Organic Gardening Experimental Farm in eastern Pennsylvania near Allentown. In beginning the rotation, we start with the hay field in fall. Soil tests are taken to establish nutrient content and fertilizer needs per acre. The amount needed usually is two tons of dolomitic limestone, two tons of rock phosphate, two tons of granite dust, plus eight to ten tons of mixed animal manure for the nitrogen. This is then applied and is ready for fall plowing. This one application lasts for five to seven years.

When using a fall plowing method, take into consideration windbreaks and degree of sloping land. Windbreaks help stop soil erosion during winter when high winds prevail and there is no snow cover to protect the soil. Compost spread after plowing helps to compensate for a lack of snow cover. The compost also helps in early spring when the heavy rains come, preventing the washing of soils on slopes. Washing can also be prevented by leaving sod strips throughout the field. The amount of strips needed is based on the degree of slope of the land. Another help is the roughness of the soil media due to the heavy root system of the type of sod being plowed under. This is also a benefit in fall plowing. It gives this matter a chance to decompose making the soil more workable for the next drop. It also makes the nitrogen content of the green manure and root system plowed down more available at a time when it is needed.

When corn planting time comes around in spring, the field has a green tint to it showing that the first crop of weeds has germinated. We then start to disc and plant. Since the first generation of weeds has been taken care of, the corn gets off to a good start. Our crop rotation begins at this time and proceeds as follows:

Soybeans then oats, wheat, rye, barley and back to the start of a new hay field.

In summing up, a good crop rotation always satisfies the farm needs from all sides. Here at the Experimental Farm we have found this rotation to work very well for our needs. But each farmer should develop what is best for his needs.
—John Keck

O. G. Experimental Farm Poultry Feed Formulas

The following are the poultry feeding mixtures in use at the Organic Gardening Experimental Farm:

Laying Mash for Chickens

Pounds:	Ingredient:
2	Meat scraps
2	Bone meal
1	Ground limestone
2	Kelp (seaweed)
2	Linseed meal
2	Fish meal
2	Brewer's yeast
2	Wheat germ

28

1	Charcoal
4	Alfalfa meal or dried hay
40	Shelled corn
15	Oats
15	Barley
10	Soybeans

Total: 100 pounds

Whole-Grain Scratch Feed

Pounds:	*Ingredients:*
35	Cracked corn
25	Barley
25	Oats
15	Sunflower seeds

Total: 100 pounds

(Note: Feed with laying mash, broadcast on litter inside poultry house, or outside on ground.)

CHAPTER 10

Organic Farms and Pesticides

IN JUNE, 1970, BOB STEFFEN presented a statement on pesticides to the Legislative Interim Committee on Environmental Problems in Lincoln, Nebraska. He made some basic points that are essential to an understanding of organic farmer's attitude to insect control methods in general, and to pesticides in particular.

The very basis of a sound agriculture is a healthy, fertile, and biologically active soil. There is a good deal of scientific evidence that insects are repelled and/or attracted by plants growing on a balanced or unbalanced soil. Certain soil conditions will also encourage fungus growth. We know that weather, poor soil structure, deficiency or excess of soil moisture or nutrients, faulty management and crop rotations all play a part in insect problems. To try and correct all these factors with insecticides on a routine basis, is, of course, not very practical. However, I wonder how many times a little better planning could have avoided insect problems. Just as the aspirin won't correct the cause of a pain, neither will killing an insect tell us why the increase of that particular species.

Twelve reports by Gustav Rohde indicate that application of a matured compost improves soil structure and has a beneficial effect on growing plant roots. Under such conditions the uptake of nitrogen from soil to plant is controlled naturally and therefore results in an optimal balance between nutrients and water, thereby resulting in a healthier plant. The crop harvested under these conditions had better-keeping quality both before and after harvest. Dr. Rohde reported that plants and soils treated with overdoses of nitrogen have a bitter taste and are more subject to insect attack.

Writes Dr. Rohde: "A characteristic of chemically oriented farming is the close connection between the application of chemical fertilizers and the need for chemical pesticides and herbicides. These two industries have advanced together and are interdependent. One of the reasons for this is that chemical fertilizers, as distinct from compost fertilizers, possess no anti-pest functions. Since they increase growth and bulk and chemical fertilizers lead to a greater water uptake in the products, this means that their cell systems are less closely integrated. This means that the tissues are more vulnerable to damage both from biological and physical sources, and therefore the keeping quality will be lowered."

There is perhaps no single factor in modern agricultural practices that encourages insect problems more than that of growing the same crop in the same field year after year. Crop rotations must once again be considered.

There is much discussion in Europe today about the modern trends to increase the size of fields by the removal of hedge rows that have been traditional with the rural landscape of almost all of Europe. There is disagreement as to the value of adapting their field sizes to modern machinery. Some conservationists feel a great loss in changing the traditional rural landscape which to them had much aesthetic appeal. Some even feel this landscape has attracted tourists to their countries. But the people of the Netherlands no longer have any doubt as to what is right. The Dutch government is paying farmers to replace field borders for various reasons. Actually, economics is the main consideration because they have proven to themselves that they get more production with fewer insect problems. A better moisture and temperature control of their fields and a more varied biological life of the community both in and above the soil together with crop rotations, avoid the extreme fluctuation of the biotic life that is found in a monoculture. Ecologists will remind us that extreme cycles of insect populations are always present in a relatively simple biological community. Here is one example where economics and good ecology are in agreement. Generally speaking, if development of a community is biologically correct it is always beautiful.

In summary then, one cannot deal with the insecticide problem adequately without:

1. Maintaining proper soil structure, and adequate fertility with a biologically active soil.
2. Encouraging a balanced fertilizer program.
3. Avoiding excesses in certain fertilizers, especially nitrogen.
4. Encouraging the use of properly treated organic matter.
5. Emphasizing production of quality crops rather than just quantity.
6. Encouraging good crop rotations.
7. Encouraging planting shelter belts and trees as well as shrubs on borders.
8. Encouraging a diversity of plant and animal life in every community.

In closing, I would respectfully ask the Committee members, before you dismiss the ideas just presented as completely impractical and out of touch with modern concepts of agricultural practices, I hasten to add that I understand your reaction.

But please address yourselves to this question also: Can we continue in the direction we are now going? Can we continue to literally pour tons of foreign material into the eco-system year after year without knowing what the long-range effects will be? Can you completely assure all of us that we are absolutely safe and have nothing to fear, either in our generation or the next?

There is an ever-growing number of people who are beginning to have some doubts.

CHAPTER 11

Alternatives to Poisonous Pesticides

A LOT OF TALK has been blared at the public lately about agriculture's need to keep on using pesticides. After all, sputter the apologists, many of the sophisticated sprays and poisons aren't as toxic or persistent as DDT — whose demise they moan over while eulogizing it as a victimized "benefactor" of mankind.

Well, what the barrage has amounted to is just that — a lot of talk. Nearly all of it consists of glib, hard-sell propaganda about chemicals being essential to turning out enough food. And with it come predictions out of a money-tinted crystal ball warning gloomily of damaged crops and higher prices. (Seems like they climb pretty fast as it is.) The truth of the matter is that insects are harder to kill than ever, and that hazardous pesticides have continued in use despite federal or state curbs. What's more, natural control techniques — many of them recent discoveries adapted successfully for farm-size application — are proving that wholesome foods can be grown in ample quantity to start catching up with the demand for vegetables, fruits, grains and meats that don't automatically deliver poison-spray residues to our tables.

Don't you believe that line about the bugs rolling over for less-dangerous pesticides. "Insects and mites are becoming harder to kill. They are developing resistance to chemicals that controlled them a few years ago," states Earl Manning in the *Progressive Farmer* of June, 1970. "One of the most resistant insects in the world has suddenly cropped up in California's San Joaquin Valley," he continues. "It's a small pasture mosquito that can withstand *1,000 times as much parathion* as is required to kill its 'normal' relatives. 'To make matters worse, this pest resists every insecticide that we have available,' say University of California researchers."

Dr. C. H. Hoffman, associate director of USDA's Entomology Research Division, states: "Resistant insects show up in nearly every crop to which insecticides are regularly applied." And around the world this problem is increasing, adds Manning. Some 225 insect and mite species have developed resistance to pesticides that once controlled them.

As for the Government's well-publicized campaign against the perils of pesticides, the series of tough-sounding edicts are proving much less effective than they first promised to be. For example, the *Wall Street Journal* reports that manufacturers' appeals to the USDA and the subsequent backlog of hearings has actually balked any real crackdown, and a check of stores indicates that sprays like DDT and weedicides such as 2, 4, 5-T are still very much on sale.

"Last November," notes the *Journal*, "the Government authorization of DDT products for house, garden and certain other uses was canceled with considerable White House fanfare. Yet three companies — Carolina Chemicals Inc., Diamond Shamrock Corp. and Lebanon Chemical Corp. — still have the right to go on marketing the chemical labeled for these very uses, while their protests await action by the Agriculture Department's tortoise-paced appeals machinery.

"These delays dismay many conservationists and consumer spokesmen. Harrison Wellford, an associate of Ralph Nader, complains that letting sales of these supposedly prohibited DDT products continue 'in effect leaves no ban at all.' "

Meanwhile, foods from standard commercial channels reflect the continuing heavy spray use. Prof. Francis Lawson, director of the USDA's Research Biological Control of Insects Laboratory in Columbia, Missouri, found he couldn't feed the insects for his experimental studies on vegetables that humans eat — because pesticides on the vegetables killed the bugs. "And we can't rear insects on ordinary lettuce, green beans and cabbage because of intolerable amounts of pesticides in them," said Lawson — who was so shocked by the experience that he now grows his own spray-free vegetables at the laboratory instead of buying them at the store.

It's no wonder informed researchers are concerned.

The solution to the pests-and-poisons problem doesn't lie in waiting another 20 years for more serious consequences to show up. Neither does it rest with foolish complacence in awaiting further chemical panaceas that wind up causing worldwide trouble and super-bugs with greater resistance or even immunity to stronger sprays. What's needed is a determined, wholehearted turn to modern biological controls — methods that nature has provided right along. These are ways to curb insects that several farsighted scientists have worked on developing for years — and ways that an increasing number of farmers have put into workable practice for large-scale food production.

Take the case of the alfalfa weevil, a pest that's been highly destructive to crops since it accidentally entered the U.S. about 20 years ago. After spraying in efforts to check it reached such proportions that tons of milk became contaminated from alfalfa dairy feed, farmers started looking into a tiny parasite, an imported European wasp with the formal name *Bathyplectes curculionis*. Much less formally, the little wasps simply clobbered the weevils — their natural food overseas — wherever they were introduced. The wasps deposit their eggs into the larvae of the alfalfa weevil, then feed off the pest host when they hatch. Agriculture Research Service entomologist M. H. Brunson at Moorestown, N.J., says a recent test indicates a 90 per cent reduction of weevils within three days. Cornell University entomologist George G. Gyrisco maintains pesticides will not be necessary within a few years (why not now?) in order to control the economic crop menace. Success against the alfalfa weevil in the Northeast has spurred hope of using parasites to curb other major pests, such as the cereal leaf beetle in the Midwest.

But more significant is what's happened with farmers. "We were being eaten up by the weevils," says John Pew, a dairyman in Burlington County, N. J. "It cost me some money the first year, when I quit spraying to give the wasps a chance to build up. But I haven't touched a sprayer since, and I'm growing good alfalfa. This has been something wonderful to see!" Entomologists simply released laboratory-reared wasps in a 30-acre alfalfa field on Pew's farm about seven years ago. Since 1957, wasp parasites have spread and been "seeded" onto Eastern farms until they are now protecting alfalfa on an estimated 60,000 acres from Massachusetts to North Carolina.

An even better-known predator who does a fast steady job of gobbling up unwanted bugs has moved from the backyard garden to farm-scale use. Ladybugs, or more technically ladybird beetles — probably the most recognizable of all beneficial insects — have come into their own as busy "farmhands." Gallon upon gallon of the hungry predators, air-shipped from the West Coast, went into Kansas fields last spring, set loose by farmers to dine on aphids, greenbugs and other pests of young milo plants. Most agreed that the "ladybug love-in" — begun when an employee of the Kansas Forestry, Fish and Game Commission and a farmer friend decided to import a supply to test their effectiveness — was a success.

One farmer, after releasing two gallons of ladybugs in a 50-acre milo field badly infested with greenbugs, reported his plants were nearly free of pests three days later. A second, who did not take part, watched ladybugs move in from an adjoining farm and rid his milo of greenbugs. One of the largest users distributed the beetles at a rate of one gallon per 15 acres at a cost of about 75 cents an acre. After viewing his successful reduction of plant pests, he asked, "Why pay $2.50 per acre to have milo sprayed when you can accomplish the job with 75 cents' worth of ladybugs?"

Other wasps have become pest fighting heroes. One of these is the almost-microscopic trichogramma, which controls the "apple worm" codling moth practically single-handed on dozens of organic orchards. It's also an effective control of fruit and vegetable insects. The wasps have been successful elsewhere, according to Mrs. Edward Horn, a former science curator at the Brooklyn Children's Museum. The Russians have released them over a million acres of croplands each year, and the Texas grower, Gothard Co., passed the 100,000-acre mark in 1967.

An old practice in peasant agriculture — alternating strips of a different crop — has gained new attention in the bug battle. "A recent finding made in Cuba may prove to be of general interest in the control of field crops pests," reports the British *New Scientist* of March 19, 1970. "Reduced incidence of fall armyworm on maize (corn), and a correspondingly increased yield, were obtained by growing the crop with sunflower in alternating strips 5.6 meters in width. There were also large reductions in the numbers of *Carpophilus* beetles in the sunflower strips, compared with unbroken areas of the crop. Some infestations were cut by more than half."

Cultivation of dissimilar crops in alternate strips is not a new practice, says the *New Scientist* article, although the Cuban investigations appear to be the first aimed at using this type of cropping arrangement specifically for pest control. The efficiency of this method, the report explains, depends on breaking or fragmenting the usual spacing of the crops, which interferes with the spread and hunting activities of the insects. "Maize is sometimes grown with cotton in Greece, and with rice in Cuba itself, and there are many other examples of long-established two-crop associations. Evidently, their primary purpose has been to increase the efficiency of land utilization. But it is possible that improved performance, resulting from diminished incidence of pests, has been one reason for the survival of these practices in peasant agriculture," it concludes.

Closer to home, farmer Wilbur Wuertz got even with the lygus bug last summer in a similar fashion. He and some University of Arizona entomologists cooked up a sneaky plot based on what they knew to be a weakness of the No. 2 cotton pest (right behind the bollworm) — that is, a preference for sweets.

Lygus bugs love to suck the sap of cotton. In the process, they weaken the plant, causing it to fail to produce or at least reducing its production. But little lygus also likes alfalfa. In fact, he prefers it to cotton. Wuertz, knowing that, planted narrow strips of alfalfa amid his rows of cotton.

The lygus bugs took the bait, steering wide of the cotton and swarming into the alfalfa. The unwitting lygus bugs were walking into the den of their natural enemies, by the hundreds — ladybugs, lacewings, and others. They love lygus bugs more than the lygus loves alfalfa.

Farmer Wuertz had put the balance of nature back into efficient operation on his Casa Grande field. He figured he saved $300 on his 50-acre test plot, at $3 an application an acre for two sprayings — which is what a similar 50-acre plot without alfalfa required to control the lygus. He also spared the lives of lygus' natural enemies.

Wuertz — who found that 2 per cent alfalfa was enough to control the bugs — planted his a month earlier than he did the cotton to have the alfalfa growing well by the time the cotton sprouted. He clipped it on the eve of the month-long lygus bug growing season, which he hoped would provide young, succulent alfalfa growth for the bugs. He's also hopeful that his "balance of nature" approach will continue to help him during the growing season.

Fast becoming another leading bug-beater is the lacewing fly. A case in point is the cotton bollworm. Southwest Research Institute in San Antonio recently turned out millions of insect eggs to raise the green lacewing (Chrysopa). The larva of the lacewing is armed with formidable pincers, and eats the bollworm and a cousin, the budworm, so voraciously that it is called the "aphid lion."

Dr. Richard L. Ridgeway, a USDA entomologist, said there is no doubt that the aphid lion can kill the bollworm and budworm as effectively as methyl para-thion, a "hot" insecticide now being used.

In a practical produce-raising example, farmer George Lindemann scattered some 750,000 eggs of the green lacewing and chalcid fly among his tomato plants to stop the ravaging tomato fruitworms and yellow-striped armyworms. For years he and his father had sprayed chemical insecticide over hundreds of acres on their farm in California's San Joaquin Valley.

As viewed by *Newsweek* (June 8, 1970), "The Lindemann family's new tactics reflect a growing awareness by U.S. farmers that biological agents can control crop pests as effectively as the miracle chemicals that have flooded the market since the end of World War II. Interestingly enough, the new trend in agricul-tural pest-control techniques is prompted less by a concern about environmental pollution than by the simple economic law of diminishing returns.

" 'We're businessmen, like any others,' says the younger Lindemann, now president of the family-owned farm. Five years ago, when confronted with a particularly severe infestation of tomato fruitworms (another name for the boll-worm), the Lindemanns sprayed their crops with carbaryl — a general-purpose insecticide. 'We killed the worms, all right,' he recalled, 'along with every other damned bug in that field.' But the worms returned later in the season, and it took repeated sprayings of the chemical to suppress them. The family spent up to $80 an acre just for pesticides on some badly infested fields. 'At the end of the year,' Lindemann said, 'we were looking at one helluvan insecticide bill.' It was then that Lindemann contracted with Louis Rund Jr., a graduate entomologist, for pest-control counseling.

"Lindemann pays Rund $7 per acre to inspect his tomato fields and $2 per acre to examine some of his other crops (melons and cotton) and to prescribe pest-control methods. By using such biological controls as the lacewings, Linde-mann currently is spending only about $5 per acre on pesticides where previously he was paying an average of $30 per acre — and saving almost $9,900 a year on chemicals. 'We're getting results,' Lindemann said. 'We had to get our costs down without sacrificing yields, and this was the way to do it.' "

One of the most active workers in the field of natural pest control is Everett J. Dietrick, whose Vitova Insectary, Inc. in Rialto, Calif., produces an astonishing number of helpful insects — especially trichogramma wasps, lacewings and a long list of less familiar parasites. Dietrick serves mostly as a consultant and "contractor" for large growers in the Southwest and down into Mexico, supplying the biological controls — which he raises in a unique "railroad-car laboratory" — and directing their timing and effective application. He's already rescued cotton and produce growers throughout the area from the brink of insect-and-economic ruin by introducing successful programs to curb the bollworm, codling moth and several other major pests.

Recently, Dietrick has concentrated attention on working out controls for various sorts of flies — particularly those attacking farm livestock and poultry. Not only do these pests limit full production of meat, dairy and poultry products, they are directly responsible for much of the undesirable toxic spraying and dust-ing around livestock.

As he explains it, the Vitova Insectary program consists of periodically releasing several species of tiny parasites in fly-breeding areas. These beneficial insects are grown in the insectary on houseflies and either mailed to the customer or released on his ranch by the entomologist, concentrating on spots where fly breeding oc-curs. The parasites attack immature flies in the manure; they puncture the pupal case to feed on the blood of the fly, and they lay their eggs in the larvae and pupae of the flies in the manure piles. The parasite eggs hatch in the immature flies, and the larvae develop by eating the flies. They grow into adult parasites, and the cycle repeats itself. This release program augments the natural biological control complex by adding more parasites to destroy more flies and thus grow more parasites.

Parasites also kill many flies that they don't lay eggs on, so that their effectiveness extends beyond the number of parasitized pupae. The stronger, more active flies attract the parasites, and are killed in greater numbers. Those weaker, inactive flies who escape parasitization lay fewer eggs than the strong flies, and transmit more weak inherited characteristics to their offspring. Thus, biological control becomes a genetic force to interact with the flies and suppress their population growth.

Dietrick, who is also president of the Association of Applied Ecologists, has put his fly-control program into successful operation at quite a few Western beef and poultry or egg ranches, now has a packaged program for farmers and even gardeners in other parts of the country. He's also at work on some predatory mites for spider mite control in strawberries and other fruits, and on gambusia fish for mosquitoes.

Insect diseases represent yet another tremendous area of no-poisons-needed control, one which has also finally been gaining the attention it merits. Dr. E. F. Knipling of the USDA, internationally known for developing the sterilization technique that's knocked out serious troublemakers like the cattle screwworm, urged entomologists to explore the possibilities of creating lethal epidemics among insects using baits spiked with germs that sicken them. The idea, he said, would be to sicken a relative few of the insects by luring them to feed on germ-packed baits — and then have them spread the diseases in epidemic fashion to hordes of their kind.

"I would like to raise the question," he said, "as to whether insect microbiologists have fully explored the possibility of finding pathogens that could be incorporated in baits to destroy insects. If pathogens were available and highly selective against each pest involved, there would be no danger of causing any significant adverse effect on the environment.

"This approach to insect control also offers the possibility of creating epidemics of the diseases among the insect population so that not only the individual insects that actually feed on the baits would be destroyed, but infections may spread to other individuals in the population to increase the level of control."

The insect-disease method has already produced several very effective controls. Among them: milky spore disease, a bacteria that infects Japanese beetles: *Bacillus thuringiensis* and nuclear polydrosis virus, a bacterial and a viral disease — both of which have shown excellent results against the cabbage looper, armyworm, diamond back moth and a number of other truck-farm crop pests. In field tests at Columbia, Missouri, for example, ARS entomologists sprayed plots with a water suspension containing nuclear polyhedrosis virus, and found that under optimum conditions, infected cabbage loopers stopped feeding within 24 hours and died within three days — M. C. Goldman

CHAPTER 12

Livestock Production and Organic Farming

TO ANYONE BORN AND RAISED in the grass-and-corn producing areas of the United States, the concept of good husbandry includes the production of meat, milk and eggs. While many American farmers are still involved in the production of livestock, the changes have been tremendous.

The production of grain has become a specialty and modern technology has made it possible to develop this monoculture without keeping any livestock around as a source of fertilizer. As grain production became mechanized, demand for milk, meat and eggs was going up so the same technology developed the production of livestock, quite apart from the land and concentrated in various areas that were determined largely by economics and not by any biological laws that we have been considering.

The ecological approach to solving all the pollution problems so much discussed today is really a rather simple one. And in doing this we may alleviate many other troublesome situations that have developed in the social, political and economic spheres.

A farm producing both crops and livestock will probably have some grass and legumes in the rotation. From the standpoint of conservation in all its aspects, this is an excellent practice. It provides good feed for the cow and even the hog and chicken can utilize a certain amount of grass and legumes. One reason the grass and hay has been omitted from the rotations on the farms is because the commercial feed lots have eliminated it from the rations of the beef animals. The production of forage has not been mechanized as rapidly as with the grains. The hay prices went up and the feed lots learned to feed cattle without the expensive and hard-to-handle forages. At least they cut the use of roughages to a bare minimum. This has brought about some interesting developments.

The formulation of the high-energy rations for all livestock, beef, dairy cows and poultry has been very rapid. Again, economics was the force behind the movement. Let us examine just a few of the results of this trend.

Dairy cows will drop in butterfat production with a decrease in roughage from their diet. Many other complications are probably the direct result of this force feeding of dairy animals.

Beef feed lots have a very high rate of liver abscess. In 1967, 9.6 per cent of the 27,859,980 cattle federally inspected were found with liver abscess. The rate is probably higher now. Researchers have concluded that this is caused by a low roughage ration. The efficiency of cattle with bad livers is cut down thus resulting in a greater loss. This problem has been alleviated somewhat by feeding antibiotics. Digestive upsets and founder are also more common in feed lots with low roughage rations.

Generally speaking the animals fed on high energy rations will have more fat in the carcasses. The percent of fat to lean will be higher. There is another aspect of this method of feeding livestock that should perhaps be mentioned here.

Some work done in England indicates that the fat from livestock fed high

37

energy ations, contains large amounts of unsaturated and mono-unsaturated short chain fats. Michael Crawford of the Nuffield Institute of Comparative Medicine has this to say about confined animals on high energy rations and free roaming herbivores:

"The high fat carcass has been developed with time, and although we may accept it as normal today, it is surprising to realize that we are now eating more than twice as much fat as protein from the domestic carcass: strictly speaking we now have 'fat production' rather than protein. On the other hand a bovid free to select its own food has a carcass which is 75% muscle and has adequate energy stores (5%).

"Such a carcass provides more than three times as much protein as depot fat." Crawford goes on to say that fat seems to replace functional tissues in animals raised on high energy feed. In wild animals one does not find this condition and Crawford asks the question "whether such lipid deposition is pathological." He further states, "The type of fat infiltrating the muscle tissue is predominantly saturated and mono-unsaturated triglyceride. Animals free to choose their own food not only supply protein but a high quality structural fat in their tissue." So the medical profession's suspicion of animal fats role in the coronary problem of man may actually need further study as to just what kind of animal fat they refer to. The organic farmers should be vitally interested in supplying this "high quality beef that can be produced on his farm without the problems just touched upon."

There is also a need to examine for a moment, the practice of feeding antibiotics to livestock. We all know the widespread use of antibiotics in the medical profession. It may in fact be used too much even in the prescription trade. But aside from this we now see the use of antibiotics on a regular basis as part of the animals' daily diet. Thousands of farm animals and poultry too, are getting the antibiotics almost every day of their lives.

Doctors are finding more and more people who are sensitive to antibiotics now and they feel the exposure is coming to their patients through the meat, milk and eggs. There is another phase of this problem that we should be more interested in. What is happening to the animals themselves and their natural resistance to disease? Are the natural antibodies that have been keeping livestock health up to the advent of antibiotics, still around? What is happening to the natural flora to the bovine digestive system or for that matter of all animals' digestive systems?

It is well known that bacteria can pass on antibiotic resistance to their progeny. In other words the germ cells do not need to be exposed to a specific antibiotic to be resistant to it. It can inherit this resistance from the parent cell which may have been exposed.

The Swann report[1], which was made in England, flatly states that transferable resistance can pass from E. Coli to salmonella bacteria. This can and I'm sure has already happened with serious financial consequences.

The Swann report urged the government to ban the use of certain antibiotics in animal feeds. They stated, "The administration of antibiotics to animals in ways at present[2] permitted — has caused harm to human health." They go on to say, "In the long run, we believe it will be more rewarding to study and to improve the methods of animal husbandry than to feed diets containing antibiotics." A more natural feed, even though the short range economic advantage may be less, may in fact be the most rewarding over the long haul.

The other phase of livestock feeding that is becoming more and more questionable in the eyes of more and more consumers is the use of hormones in livestock rations.

The most obvious and most frequently heard complaint is that of residual carry over of DES in the carcass. In a report from the Food and Drug Administration office in Washington in June, 1970[3], it was stated that of the 40 million cattle marketed last year there could have been residue present in more than 240,000 head. Official records show the residue in beef last year ranged from eight to 800 parts per million.

France, Switzerland and the Netherlands have banned the use of DES in feeding beef cattle.

One other aspect of this problem really has not been thoroughly researched and that is the question of meat quality and locker shrinkage of carcasses from animals fed hormones. The question of whether or not the gains resulting from using the hormones is indeed adding high quality protein to the carcass or just more weight in the form of fat and useless tissue has not been answered satisfactorily. — *Robert Steffen*

Footnotes:

(1) Atherosclerosis and High Energy Feed Farming Systems. *Journal of the Soil Association*, Suffolk, England. April 1970, p. 121.

(2) *Ibid.*, p. 119

(3) *Omaha World Herald*. 24 June '70.

CHAPTER 13

Producing Better Meat

VERY GRADUALLY, somewhat like the fragile grass which finds its way through asphalt and tiny cracks in concrete, the market for clean, wholesome meat, free of anti-biotics and hormones is pulling a large scale organically-grown meat production into existence. It's on the verge of creating an exciting new direct market which could well become an important alternative to the meat which most North Americans have to buy in local supermarkets.

A new efficiency is emerging where organically-grown meat producers are taking their case, figuratively speaking, directly to consumers and where they are proving first of all that there is a difference by taste.

Up in Nelson, California, in fact right on the edge of Wehah Farm, Dave and Jan Hays have been producing beef and selling directly from the small farm which they have been fixing up. Dave describes himself as a "country boy from the city of Los Angeles" and their farm is the first step in fulfilling his dream of earning his living as a food producer.

Using his little cattle farm as an innovative guide and by talking and working with his neighbors, he has worked out two concepts which could make organically-grown beef on a relatively large scale a practical possibility. The first concept he calls the "locked-in herd" which essentially means that a rancher sets aside a particular herd based upon health and history, and that from that point on the separate herd is controlled according to organically acceptable procedures (i.e., no preventive antibiotics and the like), and that a running diary is kept telling where the animals were kept day by day, specific fields, the history of the fields (relative to use of herbicides and insecticides), what type of grass, and so forth. If necessary the diary will incorporate residue tests, and procedures will require some monitoring.

The first "locked-in herd" is in process now and almost ready. The pioneer in this project, along with Dave is Elwin Roney and his son Wally, who just graduated from Cal Poly's agricultural school in San Luis Obispo. Elwin and Wally will probably be working with Dave right along, and Wally is particularly anxious to innovate and find ways to build a meat-producing industry which will restore good animal husbandry and a practical science based upon ecological principles, plus safe and sane quality food.

Let me discuss some of the advantages of the "locked-in" concept. Basically, it makes it possible for ranchers, large or small, to participate in organically-grown beef production, either entirely or in part. It's not a small consideration, considering that Dave pays a five-cent-per-pound-on-the-hoof premium for following the procedures and maintaining an accurate diary. The advantages all around provide a record of assurances which must coincide with inspections. It makes an important answer to the question: How do I know that this is organically-grown, or at least grown cleanly without using questionable and dangerous methods?

In current practices, animals are moved about and handled kind of "abruptly." Each move creates a weight loss and adds to the animals' tensions. Young immature animals are trucked into feed lots, confined in small grassless pens, stimulated and "exploded" into 1,000 to 1,100 pound chunks of fat plastic meat, and then rushed into slaughter. It is a tense situation where tense animals are made even more tense, and the system would collapse in a chaotic morass of disease if the animals were not liberally "dosed" with preventive drugs and antibiotics. (Which transfers an immunity to treatment in the human body.)

"Locked-in herds" will be held in their natural environment, in the fields, and grazing in the hills until they reach a mature 850 pounds. In this part of California, cattle are frequently moved down from the hills "old-fashioned style" in cattle drives. In a careful drive animals lose very little and they arrive at the destination in good, calm shape. The first "locked-in herd" was being brought down in a drive at the time this article was written, and it is likely that this moving method will be used whenever possible.

The second concept, phase two, is called "free-choice feeding," and it is a practical challenge to the feed lot "forced-feeding concept." "Free-choice feeding" incorporates permanent pastures and feeding bins. The animals have their choice, and may choose to graze upon fresh grass or feed upon dry feed; they can respond to natural feeding urges, and move about over a large grassy area. The concept depends upon clean, healthy animals; good housekeeping combined with optimum conditions for maintaining "tense-free" animals in a healthy environment. It is dramatically opposed to the feed lot in purpose and in operation. Feed lots were designed to produce the maximum amount of "weight" in the shortest period of time, for the least amount of money. Weight increase, utilizing stimulants such as cancer-inducing hormones, a tightly confined area, and forced feeding produces as much as four or four and a half pounds weight increase per day.

Relative to weight increase, the object of "free-choice feeding" is almost reversed, and the purpose is to "finish" the animals while *limiting* weight increases to no more than a maximum of two pounds per day. According to Dave, "any more than two pounds per day goes into fat." (There isn't enough space here to talk about the hazards of eating meat which contains excessive fat. Conscientious nutritionists are increasing their warnings, and of course pesticides are accumulated in the fat tissue.) An animal will be finished out at 900 to a maximum of 950 pounds. The goal will be to produce meat high in protein and with no more than one-fourth of an inch fat layer.

"Free-choice feeding" maintains an open, natural environment; consequently healthy animals are not easily susceptible to disease. Ample room allows adequate exercise and rest. Treatment such as the use of antibiotics is confined to "reactionary treatment"; hopefully this means calm restraint. All cattle is vaccinated against anthrax and probably for leptus and black leg, although "there is no reactionary treatment for these diseases and when the symptoms appear it's too late."

Current feeds vary with supply; at the moment only a small amount of organically grown feed is available, primarily broken or green rice rejects screened at the Wehah Farm mill. However, according to Dave, there is an adequate supply of feeds here and there which have been grown without herbicides or pesticides. Eventually Dave hopes to contract with neighboring farmers to supply organically-grown feed in sufficient year-round volumes.

For the biggest part of this year the "free-choice feeding" program will have to rely upon scattered suitable facilities, including Dave's farm, but the permanent program is planned and under negotiation now to be established on a 7,000-acre ranch owned by Don Murphy's family. Don (of the McKnight Ranch) is not a newcomer to the organic idea. A conservative, careful business farmer and a good friend of the Lundbergs of Wehah Farm, he is oriented toward developing a closer contact with consumers and maintaining the existence of the family farmer. Last year he attended the California Organic Farmers meeting in Morro Bay, and this year, under the supervision and in cooperation with Wehah Farm, he has raised 200 acres of organically-grown brown rice.

Tucked somewhere in Dave's mind is a dream of an organic packing house capable of meeting California's rigid requirements without violating organic standards. At the present time the slaughtering is done in Chico and the meat is trucked to Durham (a few miles south) to Chet Graves, owner of the Durham Locker. The meat is hung for two weeks to improve its flavor and tenderness before it is cut to order, wrapped and frozen. Chet is already moving things around to make room for the expansion, and it is easy to get the impression that he's wondering where he will eventually find enough room.

Dave's Rite-Fed Meats will be sold and delivered to consumers. He has just purchased a small refrigerated truck which he expects to be the first in a fleet of refrigerated trucks which will deliver organically-grown meat to the customer's freezer. In the first expanded phase of his marketing program his advertising will include San Francisco and the Bay area, and by the end of 1972 he hopes to reach Los Angeles on a regular basis.

All of which might give the impression that Dave is a tycoon with a pocketful of money, which is far from the fact. He is an ambitious young man who believes in the organic alternative and the opportunity which it provides. He believes that wholesome food plus profitable production sold reasonably is bound to be a winning formula. — *Floyd Allen*

CHAPTER 14

Organic Orcharding

IN 1944, A. P. THOMSON bought a 45-acre farm in the Shenandoah Valley of Virginia. The land had been literally "mined," and a great deal of work needed to be done before it could support an organic apple orchard.

Before a single apple tree was set out, a vigorous program of re-building the soil was begun. In the fall of 1945, Thomson turned under five years' accumulated growth of lespedeza, broom sage, foxtail, berry vines and assorted weeds. Over 15 tons of organic matter per acre were composted and incorporated in the soil. Several hundred tons of manure were applied. More than a hundred tons of corn fodder were composted and also added to the soil. Now, even the tree prunings are shredded and allowed to lie as compost.

In addition, Thomson applied 500 pounds per acre of ground phosphate rock, then sowed a mixture of alfalfa, sweet clover, alsike, some brome, rye grass and oats — all as a further soil-enriching cover crop. He soon found, though, that the heavy loam soil was too acid, and so in the fall of 1946 broadcast 315 tons of limestone rock. By the following spring, growth was luxuriant, much of it over lanky Thomson's head. It took hold on the farm's bare galls, and it completely stopped erosion. In October, he disked under the heavy cover, planned to use brome, oats and alfalfa as a new cover for his orchard's middle rows, and was ready to plant apple trees by December.

When it came to choosing varieties, Thomson selected first the GOLDEN DELICIOUS, a variety especially suited to his locality. Among its advantages are that it is a very high quality apple when grown well; it's a heavy bearer and a good pollinator. Another feature is that it attains an appetizing golden yellow color as it ripens. On the other hand, it hasn't found much commercial orcharding popularity because it requires a long season to mature and is one of the most winter-tender varieties. Furthermore, it is particularly susceptible to injury from chemical sprays and from mice.

Just the same, Thomson wanted to grow this variety. He wasn't going to use any poison sprays — and he was certain that the organic methods he was following would overcome the other difficulties and provide him good yields of a high-quality fruit that should be popular. Choosing this apple suggested a good name, Golden Acres Orchard. It's a fitting designation and one that aptly describes the glowing farm today. In addition, he planted RED YORK IMPERIAL, another variety well suited to the area. Both this and the GOLDEN DELICIOUS are superb for eating raw and cooking.

The trees at Golden Acres were spaced 38 feet by 40 feet apart, contrasting with the 15 feet to 22 feet spacing usual in commercial orchards. Thus, Thomson's trees are far enough apart to prevent touching at maturity, whereas the average grove becomes like a jungle. Although production per acre can't be as large, the whole orchard is benefited because the trees and fruit get better air circulation and more sunlight, and because room is provided for those enriching cover crops.

Thomson's orchard management includes another unusual feature that he calls the "fortress method." Around each tree, in a circle five or six feet in diameter and over six inches deep, he places approximately 500 pounds of half-inch dolomitic rock. What's this do? Thomson cites four major advantages:

1. It gives the trees greater anchorage against strong winds, a protection otherwise lacking where they are widely spaced.
2. It absorbs a large amount of heat from the spring sun, creating convection currents during frosty nights to provide some protection for bud and bloom.
3. It discourages mice from burrowing in around the trunks and damaging or killing the trees. (The area has an abundance of the pine and meadow mouse.)
4. As the rock "weathers," it becomes a source of calcium and magnesium for the trees' continuing requirements.

Sawdust mulching is also practiced. Thomson spreads a ton per acre each year between the rows. He uses aged Blue Ridge Mountain hard wood sawdust, selected because the rock and virgin soil of these mountains affords the trees many valuable trace elements. And, he notes, the sawdust takes up excess nitrogen produced by the legumes. In late fall, a winter hay mulch of 200 pounds per tree is applied — only to the drip line.

Currently, the entire orchard is sheet composted four or five times a year with the legumes and grasses grown between rows, not only to maintain but steadily increase the soil's fertility and structure. A rotary cutter is used to macerate this growth so that decomposition can take place quickly and a rich humus released for the trees to draw upon.

Then, too, earthworms aren't forgotten, nor their tremendous contribution to any soil-building program. Right from the start, Thomson began a sizable worm-breeding project, added them to all 45 acres and used them in compost and mulch piles. With some initial help from earthworm-specialist Dr. Thomas Barrett, the farm's earthworm population has grown to over five million per acre — all working constantly to enrich and aerate the soil.

Miscible dormant oil sprays have been found very helpful and, of course, entirely safe. Used at the right time — between the 1st of December and the end of February before any buds emerge — a 3 per cent oil solution (scalecide) go a long way toward control of scale, codling moth, leaf roller and mites, all the principal apple pests. Thorough coverage is important to get all the eggs of these insects. Thomson applies the oil when air conditions are quiet, not windy, always stirring or agitating the oil solution frequently to keep it well mixed.

For severe outbreaks of codling moth, Thomson has used Ryania, a plant extract made from the pulverized wood of a Trinidad tree. A brown butterfly-like creature in its adult stage, the codling moth's larvae are the big troublemakers, eating their way into an apple within a few hours, then leaving it to fall off the tree and to start hatching a second and then a third brood in a single season.

They first emerge about May 1st — which is when Thomson first sprays if faced with the problem. Second brood is about July 10th, then late August. He applies the Ryania at the time of each egg hatching, but discontinues it after they are brought under control. The Ryania compound, adds Thomson, is non-toxic, washes off the fruit (contrary to chemical poisons) and is effective for only about 10 days.

In addition, there's a host of predators on hand. Ladybugs and praying mantids do their share, and thousands of birds, encouraged by the multiflora rose hedges around Golden Acres, willingly pitch right in. Those hedges, by the way, also draw swarms of welcome bees who contribute to the pollinating job throughout the orchard. When some Japanese beetles showed up in 1954, Thomson put four teaspoonfuls of milky spore disease powder around each tree and hasn't had any difficulty with these troublemakers since.

Now, what about the harvest, the apples themselves? Without resorting to chemical fertilizers, hormones, defoliants, or poisonous sprays, the Thomson orchard now yields from 30 to 40 bushels per tree of top-quality, really healthful apples.

Grow a big apple orchard organically? A. P. Thomson of Bayard, Virginia, is doing it. Determination, learning and faith in natural principles have guided him. As Thomson phrases it, "We have worked long and happily with Nature, and she has taught us much. But there is, we know, an endless teaching ahead. And as it is given and received, we know also that an even better job will be done." — *M. C. Goldman*

CHAPTER 15

Places For Some Good Advice

SO MANY DIFFERENT IDEAS are cropping up, especially for the farmer who is just beginning to use organic methods. Where can you get some advice?

We strongly recommend your seeking out other organic growers whenever you can. (See the list of farmers toward the back of this publication who have answered the Organic Farm Survey.) If the editors of *Organic Gardening and Farming*, Emmaus, Pa. 18049, can be of help, give them a try.

The Extension Service of the U.S. Department of Agriculture has shown an increasing interest in making organic-type information available through its county agents. Walter John, Director of Information Services for the USDA's Extension Service, puts it this way: "With all of the concerns about chemicals in our environment, this seems to be a very opportune time for more information on fuller use of organic materials."

As part of our efforts to provide County Agents with current information, we've offered to send a free subscription of *Organic Gardening and Farming* magazine to each one as well as make our files available for their reference use. If the County Agent in your area would like the magazine, please have him write us.

And let your state experiment station know of the widespread desire to have more research work done into the organic method. It's not so much a question of who's right and who's wrong; it's simply a matter of serving a larger and larger percentage of farmers, gardeners and the American public vitally concerned with all the implications of the organic method and its total effect on the environment.

Tests are constantly underway at your state experiment station. Many of them will be of great value to you in managing your farm. A few, in fact, will be specific to your organic farm management — ones like building up the humus content of your soil with waste materials available in your area.

There's a very small patch at the Western Washington Research and Extension Center in Puyallup that hopefully will be a start to larger experiments. The Washington State University experiment, in which the plot will be managed organically, was begun in the spring of 1970. The 5-year study is under the direction of Dr. D. F. Allmendinger. The purpose of this and future experiments should be not to over-praise of over-criticize the organic method, but to provide useful information to the increasing number of farmers and gardeners who would like to use fewer materials that are dangerous to the environment.

Says Dr. Allmendinger: "There is much evidence of concern by the public in the use of agricultural chemicals, and for that reason this project takes on added emphasis."

Don't expect that experiment station information will be pro-organic — that is, that the advice will not include recommendations for pesticides and artificial fertilizers. But a great deal will be most useful.

At least once a year, a booklet listing all publications prepared by the state experiment station is available to state residents. A postcard to the station will put your name on the list to receive it — as well as other special booklets. A call

46

to your county agent will also get you this information.

Which varieties grow best in your area, local weather and soil conditions, advice on cultivating, suggestions for new plantings — these are important facts you will get.

To be successful, projects — no matter how small — require the same basic economics as a large-scale operation. Simple methods of keeping books should be developed. One of the most important tools is the lead pencil applied to paper with firmness — for figuring costs, labor hired, as well as the number of feet or yards between rows so that machinery can be properly adjusted. You should know the spacing necessary between plantings to judge the correct amount for feeding your family, livestock, poultry, etc. You should know the expenses involved with each crop. If it is a business, this is essential; if it is for pleasure, more or less, it may help you decide whether some other crop would be more to your liking — and advantage. Farm Account and Social Security record books are available at a minimal cost from the stations.

Check the problems that appeared in previous growing seasons and learn how to avoid the same ones in the coming season. Give thought to some of the newer varieties of seeds for vegetables, grains, grasses, flowers, etc. Are some better suited to areas other than yours? Check this out with the leaflets from your county agricultural agent's office. Most states offer a circular on fresh market and home-garden vegetable varieties. Try a few new ones and follow planting and soil factors.

Under the heading of Farm Management, state publications offer topics dealing with:

"Getting Started in Farming."

"Where and How to Get a Farm."

"Part-time Farming."

"Managing Farm Finances."

Extension Service and research publications about agriculture and its many phases that are published by your own state deal with conditions covering local problems and remedies for them. This is the main reason for getting information from your own area.

The reports that emanate from the Department of Home Economics fascinate me. These reports are geared to the distaff side of the family. There are ideas galore on new methods or suggestions for running a household and coping with numerous irritabilities — everything from food recipes your family may not like to rearranging the furniture and misbehaving children. Corrective measures are invariably offered.

Information is provided for purchasing the best type of bedding, clothing, textiles, their care and use. Food-buying, food-canning and preservation and various methods of cooking, practical or dress-up dishes, party-giving, and other things that tend to make a happy home are offered free or for a very small amount. These experts are full of fresh ideas. A few interested women in any township may ask the Home Economist to set up a program for a project of interest to that particular group. Lessons are offered in upholstery, making of draperies, slip-covers, simple sewing, or more complex techniques. You ask for it — and it usually is arranged.

All county agricultural agents arrange meetings, lectures, demonstrations on tree pruning, bee-keeping, cattle feeding, raising tree seedlings, etc. If residents in their area are keen about getting more information and exchanging ideas with the neighbors, with teachers, college professors and practical businessmen, the county agent goes into action. If you are interested, request that your name be placed on the mailing list.

The Food Business Institute of the University of Delaware in Newark, in cooperation with Federal Extension Service of the U.S. Department of Agriculture, published a 118-page report in 1965 on Farm Roadside Marketing in the United States. This is another fascinating study for the serious market-stand operator. Cost is $2.00 from the University of Delaware, Newark, Delaware.

The 4-H Clubs — cooperatively sponsored by the federal government and the states through the county agents — represent another important service, especially significant since it concerns itself with young people. Many a youngster in the field of agriculture today started in a 4-H group when he was about 10 years old. Here he learned how to handle stock, feeding and grooming, and showing at county fairs. As he grew older and continued with his program, new horizons opened up. With the aid of teachers and his local county agent, and the interest of his parents, he entered state fairs and if outstanding in his field, strived for national ones. Later the FFA carried on more detailed and heavier programs. By this time the youngster, whether male or female, achieved the art of self-discipline, patience, cooperation — or how else could he manage his animal on the show ground. To me, it is always a delight to see youngsters and their animals at fairs. There are now more than 25 different categories offered to boys and girls wishing to take part in a 4-H program. In my area alone, there are a half a dozen or more of these youngsters who now have their own dairy herds — started from one calf won as a prize in a 4-H program.

When I first started farming, I listened to some of my knowledgeable neighbors and bought a family cow and a few pigs. The county agent soon set me straight by convincing me that an ordinary cow ate just as much feed and needed just as much care as a good producing cow, who would return more money by heavier milk production and worthier calves. It made sense, and I soon switched and bought my foundation Guernsey calves from an accredited herd. As my herd grew, my agent became more helpful, and each suggestion paid off for me.

I didn't see how it could be accomplished, but by following the advice of my county agent for both cows and hogs we developed a successful program. Later I learned that this particular agent was known throughout the state for his outstanding work with 4-H youngsters in cattle-raising projects.

As a rule, the office of the county agricultural agent or field man is located in a U.S. government building in the county seat. The agent is a very busy person, so it would be sensible to phone his office or write a note before visiting. His telephone is listed under your county's name, Agricultural Extension Service Office. Your local library can tell you where to call if you have any difficulties.—*Eva Wolf*

PART III

ORGANIC FOOD MARKETING

CHAPTER 16

What Makes A Food Organic?

ALL THE TALK about natural foods . . . health foods . . . organic foods . . . over the last several months has beamed a spotlight on a basic point: *Just what makes a food organic?* The need to determine clear-cut, satisfactory standards as a measure of quality represents one major first-order-of business objective as the boom roars on.

Today, of course, the general concept of organic food is well understood by the public. There's no quibbling about the words or the intent when *Life* magazine's cover, for instance, headlines its colorful feature "Organic Food: New and Natural." Inside the superbly photographed and reported story on "the move to eat natural," the swing to organic foods comes across with impact both visual and editorial.

"The ideas are simple and appealing: we eat too much, mostly of the wrong things; our food comes to us not as nature intended, but altered by man during both growth and processing; and this tampering has produced increasingly bad effects on man and the ecology . . ." So starts *Life*'s article of Dec. 11, continuing with a definition that rings pretty clear: "True devotees not only reject the instant world of brown-and-serve. They also insist, and will go to great lengths to insure, that all their food be grown organically, that is, without artificial help of any kind: no chemical fertilizers, no pesticides that linger in the soil. Meat must come from animals raised without benefit of antibiotics. Foods should be free of what they consider dubious chemical additives, whether for color or flavor or preservation."

Some of the earliest definitions — those dating back to the 1940's and 50's or before — include fundamental points that remain constant and important. At the same time, developments in growing techniques, the whole range of agricultural materials and equipment, as well as the massive food-producing field, have expanded so vastly that more gauges to "organic quality" have become imperative. There's been a floodtide of new evidence about vital nutritional factors, along with increasing discoveries of the hazards and outright disastrous effects of both pesticides and food additives. There's also been a sharp upswing in the soil aspects of environmental concern — in progressive steps toward recycling more of all types of wastes and in building or maintaining fertile, ecologically-sound soils as a part of such programs. Then too, there have been remarkably long strides made in the control of insects via biological methods and other non-chemical means. Put together, these factors add up to a mandate for an unequivocal standard based on *today* — when so many people are turning hopefully and eagerly toward organic living.

One of the original concise defintions of organic food appeard in *Organic Gardening* and *Prevention* magazines in 1953 under the heading: "What Does Organically-Grown Really Mean?" The paragraph stated: " 'Organically-grown,' when accurately applied in description of foods and crops, means specifically that these have been raised on soil fertilized by organic methods only. It particularly indicates that no chemical fertilizers, conditioners, insecticides or any such type of spray, pesticide or preservative has been used at any time in the growing or preparation of these products. A soil receiving the full organic treat-

51

ment is not deficient in any element. Any deficiencies that do occur are corrected by natural fertilizers."

It's important to underscore the fact that growing foods organically *does not* imply merely *avoiding* toxic chemicals, artificial fertilizers or sprays. There is a significant *positive* aspect to the whole idea — that is, maintaining a naturally rich, productive soil by making sure it *receives* the desirable organic and rock-mineral materials. It means that *positive* steps *are* taken to encourage natural insect control — a balance of beneficial species, birds, cultural techniques, etc. As vital to the organic idea as keeping away from poisons is, the negative approach alone is misleading. Everything from recycled garbage and leaves to lady-bugs and purple martins represents a part of the whole concept. — *M. C. Goldman*

CHAPTER 17

Making It On Fewer Acres

IS ORGANIC FOOD RAISING a paying proposition?

We think it is or can be made so. Public demand for organic produce is steadily rising. Meanwhile the growers are becoming more efficient with experience; they are learning to diversify their crops in order to satisfy their market, and also to extend the season at both ends. "We grow three crops a year — very early, conventional and very late," Paul De Fere tells you. Finally, the growers are reaching out after their legitimate market, building roadstands, distribution centers and warehouses to make it easier to get fresh, ripe fruits and vegetables into the home.

We have found on investigation that these so-called "little" men can hold their own with the big commercial outfits weighted down with poisonous chemicals, expensively heavy equipment — and equally heavy mortgages.

"I couldn't manage this small orchard if I had to keep paying for poison sprays," Leslie Barnes told us in Belding, Michigan.

Yes, there is room for the small producer who is willing to work to reach the specialized market that's waiting patiently for him — the country's organic-minded consumers. Let's take a quick look at the record, going up the scale, acre by acre.

Out in Lake Villa, Illinois, about 25 miles above Chicago, Howard and Elsie Ensign manage to raise just about everything on their 10 acres of well-cultivated land.

Yes, the Ensign homestead is a prime example of subsistence farming that is competing successfully in the open market with commercially raised foodstuffs. He is a consistent composter who makes full use of all crop residues, manure from the goats and cows. In addition to raising his own hay, he devotes two acres to open-pollinated corn. Among his cash crops are carrots, lettuce, cabbage, green beans, tomatoes, eggplant and comfrey.

In one year, the Ensign beehives produce 1,000 pounds of honey — another source of income. At Easter, chevron or young goat's meat is a particularly good "cash crop." Then there are the strawberry and raspberry patches, which run by Mrs. Ensign, bring customers in from miles around. In fact the seasonal berries are responsible for added sales to the "organic shoppers" who seem unable to resist comfrey. So the Ensigns are considering a root cellar to extend their selling season.

Wilbert Walker's 15 acres are situated on a back road near Perkasie in eastern Pennsylvania. But folks drive all the way up from Philadelphia — 35 miles to the south — in season to buy his late-ripening strawberries. He sells five to six thousand quarts well into July, and could sell more. About 200 boxes are picked each day by school kids who are paid 15 cents per quart. Later in the season, the shoppers fill their own boxes. Most of the berries are sold right out in the field at 50 to 70 cents per quart.

Cantaloupes are another big-selling crop that brings customers up from Philadelphia and outlying areas. This amounts to repeat business because the 'loupes come in late August and early September, and Walker has found, like many other organic growers, that most folks don't just buy one or two items. "They shop

53

around, and that's what we want," he reports. To stimulate sales and get top flavor, he leaves the big fruit on the vines until they reach full maturity. This means less spoilage but more inspection, so when the ripe cantaloupes are picked they "separate readily and cleanly from the stem."

But it is just such "custom-made" attention that makes organic produce so much more desirable than the fruits and vegetables at the supermarket.

Walker also supplies his customers with the staples — corn, tomatoes, potatoes, beans, greens — as well as comfrey and canned tomatoes and juices. Business has been picking up steadily in recent years and, despite his backroad location, he is now expanding into a second patch with more acreage plus a handsome combined stand and showroom.

It was a cold February day when we visited Paul De Fere's flat 20-acre growing patch in Bay Shore, Long Island. But we knew that his Brightwater Farms harvests are delivered daily in season to the Good Earth Natural Food Store on First Avenue in New York. It's a haul of about 40 miles from the farm to the market.

Brightwater has grown from a 60-by-300 clearing to its present size in 15 years, which means that De Fere has found that growing organic foods can be made to pay. The sign next to his roadside stand — just off the Sunrise Highway — reads: "Eat Unsprayed Foods."

Since demand is continuous over in New York De Fere grows three crops a year — "very early, conventional and late." He reports that "we plan for it, and have 17,000 strawberry plants growing at the peak of the season." Other organic staples include early onions, corn, potatoes, tomatoes and cabbage. Like other growers he depends on repeat orders and has found that satisfied customers buy widely, from eggs, through fruits to vegetables, honey and juices. His stand is open for business from May through December, and he periodically prints up a list of available foods. That is what 20 acres, plus careful management and planning have done at Brightwater.

"I couldn't manage this small orchard if I had to keep paying for poison sprays." That's what Leslie Barnes tells you out on his 50-acre place in Beldin, Michigan — not far from Grand Rapids.

"I'd have to have at least three times the number of trees I now have," he continues. "The organic method definitely pays out for me — the number of people who want unpoisoned fruits is increasing all the time."

Barnes ships apples, peaches and peas to just about every state in the union. He reports increased crops while the topsoil on his land has thickened 1½ inches in 15 years, thanks to his foresighted policy of continuous and heavy mulching.

In addition to interstate shipments, he delivers truckloads of fruit to an organic cooperative in Minnesota. He charges the same price for his apples and pears as his neighbors, but gets about 25 cents a bushel more for his peaches. An optimist, but practical, Barnes points to the time when there will be a chain of central distribution points where increasing numbers of organic growers can deliver produce for storage and marketing.

Howard Dull's story of how organic soil can out produce chemically-treated land just about clinches the case for how organic living can be made to pay. His 100-acre farm lies halfway between Dayton, Ohio, and the Indiana line, along Route 49, at Arcanum. He hasn't always been organic, but he is now.

Today Howard and Olive Dull are doing all right. In addition to a pair of scholarship-winning youngsters, they're raising 70-bushel wheat, Toggenburg goats, Holsteins, Angus and Herefords, 150 fryers, corn, potatoes and soybeans. When we called, Olive Dull served us a "scratch-up" midday country meal of chicken, potatoes, corn, asparagus, lots of fresh salad greens and home-baked bread. Everything was organic, and everything has been grown on the place.

"The grocery bill for the four of us was $8 this week at the store," she told us. We also knew she had just canned 223 quarts of tomatoes.

The story of how organic soil can out produce chemically-treated land is just as simple and direct. "Everything worked out just right," Howard Dull says.

"The weather came on good; we had plenty of rain when it was needed, and over at the feed mill where they sell the chemical fertilizers they were talking about 60-bushel wheat. They wanted me to get in and take some of the fertilizer. But I turned them down — I wanted to stay organic."

So while his neighbors were using chemicals and talking up their 60-bushel wheat, Howard Dull kept on with his granite and poultry manure. Finally he called in the County Agent. "I think I've got 60 bushel wheat raised organically," he said.

The County Agent took one look. "I think you've got better than that," he said, and it turned out he was right. At the end of the season Howard Dull's organic wheat stood at 70 bushels to the acre — a record in a record breaking county!

Howard is just as progressive in his marketing as he is careful in producing top-quality foods. He's trying to build a local market for his organic beef. Since out-of-state shipping is complicated, he's reaching out locally, to nearby Cincinnati and Dayton, in search of customers for both his meat and produce.

"They're talking about building a distribution center in Cincinnati," he told us, "as a warehouse to take our stuff, but it's slow in coming."

Organic farming is easier today. That's what Joe and Ted Carsten tell you at their 200-acre Deer Valley Farm in Guilford, New York. "We do the organic soil-building job with less work and in less time," they stress. That's encouraging news in these times of soaring costs and prices. Their system calls for composting more than 400 tons of manure a year and spreading it with mineral additives over their fields. During the winter, it's hauled to where it will be spread, and turned several times in the spring before being applied.

Although extra work is involved, the Carstens are convinced that the program pays for itself in the long run. Yields have steadily increased, and fertilizer costs are down, thanks to the composting program.

The Carstens have developed a mail-order market for their flour and non-perishable items. They specialize in a hard wheat — KAN-KING — that assays at 14 percent protein, excellent for bread. They have about 13 acres in wheat which supplies flour for the combined bakery-organic food store in nearby Norwich.

The Carstens also have worked up a delivery route that takes in Syracuse, Ithaca and Albany, while organic cooperatives furnish another outlet for their produce in New Jersey. — *Maurice Franz*

CHAPTER 18

Organic Farmer Certification Program

IN AN EFFORT to clear away the cobwebs and throw out the monkey wrenches, Rodale Press through *Organic Gardening and Farming* has initiated, first in California, a program for confirming the farming methods and material of organic farmers. Essentially, it is a simple, modest program designed primarily to help organic farmers identify themselves for the benefit of organic food buyers. The program incorporates certain very minimum standards which are fundamental to good, basic organic farming methods.

The heart of the program is a farmer's commitment to rebuild the soil, and each farmer has committed himself to either maintain a three percent minimum humus content or to build a three percent humus content within five years. The program emphasizes the family farm. Each farmer is encouraged to give his farm a name which will also be a brand name. Organic seals will be made available which identify the farmer's commitment, his commodity, and his location. In principle and in fact, committed organic growers are saying, "We have nothing to hide. Our farms and our methods are open for public inspection."

The program incorporates soil analysis, certain residue analyses, and personal inspections. When I visit a farm I look for evidence that there is a commitment to basic organic farming methods. How are weeds controlled; can I see weeds where there should be weeds? I look for birds and insects and other evidence that there has been no recent spraying or dusting with toxic insecticides. If I see spraying equipment sitting around somewhere, then I like to see where the dormant oil is kept. Timing is an important part of the program; that is, understanding when certain commodities are sprayed with what, and then visiting the farm at that time and perhaps taking samples for analysis — all of which provides information that helps the farmer to verify his commitment.

The program is not foolproof; someone could use a little urea or 5-10-5 and go on for awhile without detection, provided that he is also following procedures to rebuild his soil. What we can do this year will not be as complete as what we can do next year. The soil and residue analyses available to us now are not as thorough as those we hope to have in the near future.

In the main, the program has already acted upon farmers and distributors, and there has been a general upgrading in organic quality and renewed interest in strengthening minimum standards. A few national distributors have indicated that they are now working on a dovetailing program which will include farm identification. — *Floyd Allen*

Organic Farmer

Certified by Organic Gardening and Farming Magazine

THIS SEAL MEANS what it says: the farmer is organic. He is using recognized organic growing practices to raise his crops. The seal and certification have already been awarded to many farms in California.

WHAT IS ORGANICALLY-GROWN FOOD?

Organically-grown food is food grown without pesticides; grown without artificial fertilizers; grown in soil whose humus content is increased by the additions of organic matter; grown in soil whose mineral content is increased with applications of natural mineral fertilizers; has not been treated with preservatives, hormones, antibiotics, etc.

THAT DEFINITION is the heart of the **Organic Farmer** program. It spells out clearly the way his food is grown and produced. It defines the methods used in raising wholesome food, not in processing or manufacturing anything else.

Display cards with this definition and the **Organic Farmer** seal are being sent to over 3,000 health food stores, shops, stands and markets. It's part of OGF's program to get grower and public together—right.

CHAPTER 19

National Farm Organization Marketing Program

ANOTHER MAJOR STEP toward increasing the number of organic farmers in the U.S. has been taken by the National Farmers Organization (NFO) — whose sole objective since its creation in 1955 is to secure better farm prices through an aggressive program of collective bargaining.

There is nothing particularly "organic" about NFO — that is, it has no commitment against pesticides or commercial fertilizer use or to rotation programs that would build up humus content. NFO is a marketing organization, and a tour of its Iowa headquarters shows how effective it can be.

Yet NFO does have some "organic" qualities — a basic part of its crusading, anti-establishment spirit is designed to make it profitable to be a family farmer and to stay on the land. And like the organic and environmental movement in this country, NFO is controversial, raises the emotions, and is criticized for its actions while praised for its goals.

The NFO marketing program for organically-grown crops was sparked by Vincent Spader, assistant director of the NFO grain market division, who brought everyone together. His first contact with the organic market came about a year ago when NFO members in Minnesota who were long-time organic farmers urged NFO to market their grain crop to the natural foods market. This August, NFO decided to set up a separate marketing route for organically-grown crops — the final push coming from Michael Scully of Buffalo, Illinois who told Spader he would join NFO if it would market organic crops. Scully also agreed to help set up certification standards and then contacted Barry Commoner, internationally-known ecologist from Washington University to get his views. Kevin Shea, scientific editor of *Environment Magazine* and an associate of Dr. Commoner, researched the problems of organic food certification prior to the meeting.

Spader was realistic in describing the NFO organic market program.

"We don't plan to head in all directions. We'll start with one commodity (grains), set standards and go from there," Spader explained. "A lot of our members have been pushing us to do something. We intend to set up standards and organize a system of moving organic foods into the market. We welcome comments on how to do the job better."

As presently planned, NFO members who want to have their crops sold as organically-grown must sign a Natural Foods Agreement, specifying the kind of grain, class or variety, grade, quantity and "For Sale After-Date." The agreement also requires the farmer to list chemicals, fertilizers, organic products, rotation schedules, fumigants used during the past 5 years.

Dr. Shea outlined the laboratory tests required to certify both the farmer and his crop. The tests at this point would evaluate levels of "unnatural contaminants" — first in soils before the farmer planted his crops, and then in the crop itself before marketing. Each load would be tested as it moves to market. NFO also plans to have the right (in contract) to inspect and test NFO organic crops in stores as a further way of extending certification through the distributor level.

Spader envisions the creation of an NFO Natural Foods Block. "Anything in that block must meet standards set by NFO regarding quality and being toxin-free. The Block will mean a steady supply available for distributors of organically-grown foods certification standards, and (most important to NFO) a firm price to the participating organic farmers."

Kevin Shea is an objective and a most interested observer of the increasing market for organically-grown foods. But like other objective observers, he foresees the pressures for controls and safeguards that must eventually come from the U.S. Department of Agriculture, the Federal Trade Commission and the Justice Department. Others sense the tremendous pressures that will come from the agri-chemical companies and the commercial food industry. One well-known marketing authority speculates on the "inevitability of a scandal exploding in organic food. Logic leaves no alternative conclusion other than that a product line enjoying high profit margins, whose source cannot be readily identified, and which has moved into controls too little, too late, simply must include a large degree of malpractice."

Considering all the "scandals" that have been making headlines almost daily on the toxic substances discovered in commercial foods bought in supermarkets, there most likely will be cases of malpractice in the marketing of organic foods. But what is absolutely essential now is to develop a monitoring program that incorporates the best of the information and methods available. We should not be defensive, paranoid or holier-than-thou in explaining the standards used to certify organically-grown foods. By doing the right things now, says Kevin Shea, "when the crunch comes, you will be totally protected."

In setting up its Natural Foods Block for 1972 (whether grains or any other crop), the NFO is using the definition for organically-grown which has appeared in *Organic Gardening and Farming*: "food grown without pesticides; grown without artificial fertilizers; grown in soil whose humus content is increased by the additions of organic matter; grown in soil whose mineral content is increased with applications of natural mineral fertilizers; has not been treated with preservatives, hormones, antibiotics, etc."

Concludes Vince Spader: "We believe a good many producers within NFO are qualified to market their harvests as organically-grown. We know it will take time to build a solid foundation and we don't expect miracles." — *Jerome Goldstein*

CHAPTER 20

Forming A Producer's Cooperative

ONE OF THE MOST PROMISING avenues for marketing organically-produced foods is the cooperative. The farm cooperative is nothing new to American agriculture, but it's new to organic farming. One of the first organic co-ops started several years ago in San Antonio, Texas, banding large-scale gardeners and small-scale farmers together to ease mutual marketing problems. More recently, an already established co-op turned to organic foods and a Canadian economic agency studied the potential for marketing organic foods through a regional farm cooperative.

The last is only a paper cooperative; it exists only in the minds of a few economic researchers and planners. But it stands as an indication that the Canadian government is willing, at least, to recognize the potential of the organic foods market and the viability of the family farm within that market. Its basis is a research report by the Agricultural Economics Research Council of Canada.

The report, titled "Potentials of the Organic Food Market and Implications to Gaspe Farmers," is the consummation of a lengthy study and analysis of the organic foods market in New York City, Boston, Montreal and Toronto. "The future outlook of the organic food market appears highly favourable," the report states. "Besides retailers' and wholesalers' expectations of increased sales volumes in the next few years, a report from Barron's *National Business and Financial Weekly* estimates that by 1975 organic food will hold 40 per cent of the total food market."

The report continues: "While the demand for organic foods may exist, there is no guarantee a new organic farmer will make larger profits." Profits, says the report, "will depend largely on how organic growers and merchants organize themselves and establish a system of certification with standards of quality. An organic food association seems the preferred organization . . ."

These conclusions are then related to the Gaspe situation. The potential is good.

The Gaspe, situated in the Province of Quebec, is a 200-mile-long peninsula jutting into the Atlantic Ocean at the mouth of the St. Lawrence River. It is geologically, historically and scenically rich and economically poor. Its agriculture is small-scale and diverse. And it is, unconsciously, organic.

"So far as you can discover," writes frequent summertime visitor Ruth C. Adams, "chemicals are almost unknown — have been unknown down through the centuries. Huge piles of manure by each dairy barn are spread lavishly over the fields. And that's the fertilizer. Never have pesticides or chemical fertilizers defiled the soil, or most of the soil, in all of Gaspe. Instead, the time-honored methods of agriculture have prevailed."

According to the Canadian report, the Gaspe offers "qualities of cleanliness and isolation, occurring not by design but by historical accident — poverty. The area is capable of growing many organic products which are in high demand in both U.S. and Canadian markets. Many people in organic food marketing in both Canada and the U.S. have a favorable image of the Gaspe as a clean food producer, already suggesting that an Organic Gaspe product could achieve a premium label status for its organic food."

Citing a number of requirements for exploiting the market, the report offers both short-term and long-term strategies for doing so. The former would function until a permanent, long-term agency could be formed.

"The Gaspe farmers cannot individually exploit the organic food market as effectively as they can as a group," the report says. "An agency is needed to provide certification, technical information, acquire seed, contact buyers, make contracts and to package and ship products. The ideal group would be one owned by participating farmers . . . The staff of such an agency must be free to play the entrepreneurial role without undue restrictions. One solution would be to form a Gaspe Organic Food Growers' Association, which would certify products as organically-grown and have a separate marketing agency or agent who would purchase, process, package and merchandise all organic products produced in the area under one label. This approach would separate the certification and sales functions," continues the report, noting that "this separation appears very desirable in achieving consumer acceptance."

The report suggests that such a program would easily enable the Gaspe farmers to market their produce, but acknowledges that it would take a year or two to get established. Thus, a small-scale trial program is suggested, "in which a small number of farmers may be assisted to try organic farming . . . It is very important to keep profit expectations at a moderate level and to provide considerable morale and, if necessary, economic support. If the farmers have a satisfactory experience the first year, they will continue and their neighbors will follow their lead."

Thus, the report figures it can be done; but only time will tell whether an organic producers' cooperative can and will come to fruition. The potential exists, and the need exists.

In pockets and regions all over the United States, similar needs—for viable marketing programs to enable poverty-stricken family farmers to help themselves —exist. In Virginia, a cooperative marketing program similar to that outlined in the Canadian report is coming to fruition. The sale of 40 acres of organically-grown cucumbers generated $12,000 for the Southern Agricultural Association of Virginia, Inc., and prompted the group to commit another 60 acres to organiculture and to consider devoting the rest of its acreage, too. It just may be economic salvation for the poor, black farmers the state and federal governments ignore with their programs.

Many small black farmers or sharecroppers are being forced into the ghettos by chemical agricultural giants. But the members of the Virginia cooperative are finding out that they can stay on the land and live. One member described harvest from the 40 organic acres as bigger and far superior to their chemical acres, and better able to withstand drought. If ghetto residents could only reach the land again, to nourish it and themselves back to health . . .

This is the hope of the Rural Advancement Fund, Inc., of the National Sharecroppers Fund—and so far its plans are succeeding on a small scale, mostly through programs aimed at keeping people on the land.

"Throughout the South," says Jim Pierce, the NSF's executive director, "small farmers, tenants, and sharecroppers are changing from row crops to specialty crop production, marketed through their own cooperatives, and substantially increasing their incomes."

Fay Bennett, director of development for the Rural Advancement Fund (1346 Connecticut Avenue, N.W., Washington, D.C. 20036), believes that organic is the way many of these developing cooperatives are headed.

"This year has been a learning year. The local farmers now have a good idea of what we mean by 'organic.' They know what went into the soil and they produced the most beautiful, best-tasting, longest-keeping cucumbers and squash anyone had ever seen in his life. We believe that most of them are convinced that

61

organic is how they want to go next year.

"This whole program and the one in Burke County, Georgia, are bringing new hope and new life to two rural areas which heretofore seemed doomed to go down the drain. For years they had faced the attrition which all small farmers had faced and, in addition, they had faced discrimination because of race. They were planting row crops (tobacco, cotton, corn) as their fathers and grandfathers had done before them. They had not learned about diversification and they could not afford modern machinery, irrigation equipment, technical help. Our organization supplied these essential ingredients. We believe these communities are now turned around and are on the way up. They are developing health programs, discussing nutrition, housing, sanitation and other things needed for a good life."

In addition to marketing produce (grading and shipping), the co-op also provides some tractors and equipment, a credit union, a store where charity items (clothing) are given away or sold at a nominal fee for those who can pay. It is also a place to meet.

Although the co-op has done wonders, and is viable, there are huge problems. Organization and leadership are crucial and must come from within the community to be accepted. Money is often viewed as a cure-all, and all problems and solutions are related to costs.

Charles Dixon, a member who functions as a field representative for NSF, is a strong force behind the co-op. He said, "The original idea was to teach the farmers to do better what they were already doing. This could and should have been done by the county agents and the state colleges, but they wrote off the people we're working with a long time ago."

What the farmers were already doing was a form of subsistence farming, with tobacco as the cash crop.

Effort was made to expand plantations of vegetable crops rather than tobacco, and soon it was found that cucumbers, squash, and peppers seemed to be most promising alternatives. The first year 150 acres were planted, and participation was enthusiastic at first, but waned as the crops matured with no markets available. A large portion of the crop rotted on the vine.

The next year a few markets were located, but the disappointed farmers failed to respond by growing crops. 1971 found increased crop production with adequate markets available, but still, as a sign outside says, "We ain't no big thing, but we're growing."

The fundamental marketing problem has been satisfaction of the volume demands of the large processors. Contracts can be entered to provide large volumes of produce, but contracts for smaller volumes aren't as easy to come by. And in the middle of it all, the growers are suddenly becoming anxious to try organic farming, which is a different market altogether.

One hundred experimental acres will be planted to various organic crops in 1972. At present Dixon anticipates growing cucumbers, squash, field tomatoes, carrots, beets, early peas, string beans, sweet corn, and white and sweet potatoes. If funds are available he would like to plant another 100 commercial organic acres. After results from the specific crops are in for 1972, recommendations will be made to farmers growing organic crops.

Although no recommendations have been made as yet, there appears to be little difficulty getting the farmers interested in producing vegetables organically. "I would like to emphasize that those 40 acres we grew organically this year outproduced the best of our chemical plots in spite of a later planting for the organic portion," Dixon said.

On the basis of these initial results, however, it would seem that organic is the way for the small-scale farmer to go and that a co-op is his best marketing outlet. Two very important elements of the organic co-op—the certification of members and production standards—haven't been confronted as yet by SAAV, but they were initial issues faced by the Family Farms Organic Growers' Association of San Antonio, Texas.

The members of this pioneering co-op got together because they were looking for markets for their organically-raised produce. A primary purpose of the association is to promote its brand name, which all members in good standing can use to make the sale of their organic food easier. The founders of FFOGA are large-scale gardeners and part-time farmers who sell their produce in season.

The Family Farms approach to solving the organic food problem is the brain child of Malcom Beck (P.O. Box 20318, San Antonio, Texas 78220). Beck was delivering some of his excellent organic tomatoes to a drive-in grocery when he was struck by the thought that the job of explaining organic foods to storekeepers would be easier if his tomatoes had a brand name. He thought of those little stick-on labels which have been used to upgrade consumer opinion of bananas, cantaloupes and other fruits and vegetables. Small labels which proclaimed the food as organically-grown would simplify Malcolm's marketing problem no end. They would insure that both the storekeeper and the consumer be continually reminded that those tomatoes were organic. And people who were out looking for organic food would be able to find it easily.

Then he thought that the labels would mean even more if they were on a wide range of organic food available in all parts of the San Antonio market area. He thought of his friends who were also organic gardeners, and who could grow different foods. One man could grow organic sweet potatoes. Another had a producing orchard. The idea of a cooperative quickly came to mind, all joined under the banner of one trade name they could promote together.

After several months of discussion and planning, the cooperative was formed. There were some problems to be solved, however. Deciding on the brand Family Farms Organic Foods wasn't difficult, and that name was copyrighted.

But standards had to be established and were. They are to assure that what is touted as organic is indeed organic.

Enforcement of the organic standards is a matter of high priority among the Family Farms associates. A committee has been appointed to be sure that the standards are applied by all members. Their approach is constructive. Instead of trying to snoop on members, the enforcement committee visits gardens and farms with the idea of trying to help solve growing and marketing problems.

The Family Farms group is still young, but their concept of an organic food producers' cooperative seems to be passing the test of actual operation. Other organic growers should know what the people in San Antonio are doing. And what SAAV is doing and what the Canadian economists are suggesting for the Gaspe. Growers should think up their own brand name and set standards for the organic food that will carry that brand. They should start their own producers' cooperative for organic food growing and merchandising.

It is an idea with plenty of potential, mainly because it gets local people working together to solve their own problems. Organic farming can build strongly on a base of well-managed cooperatives.

CHAPTER 21

Direct-To-Consumer Marketing

MANY NEW DEVELOPMENTS are taking place in the organic foods market. No one has all the answers, and yesterday's answers may not be correct today. And tomorrow more changes will occur.

In the early 1900's in India, an Englishman by the name of Sir Albert Howard began to define a method of growing crops that produced healthier animals and healthier humans. In 1942, J. I. Rodale started a magazine called *Organic Farming and Gardening* and the idea somehow managed to survive.

It used to be that most people had to be sick in order to care about survival. But that's not true in the 1970's. Young people care about survival along with older ones. Republicans care, and so do Deomcrats. Short hair, long hair—skirts or slacks—boots or sandals—suits or dungarees . . . people care. There are millions of them who care about the food you grow—and *how* you grow it.

How many? It's difficult to put accurate figures on paper. In January, 1971, *Organic Gardening and Farming* magazine had 600,000 subscribers. A sister publication, PREVENTION, has almost 850,000 subscribers. That adds up to about 1½ million families or about 6 million people. Let's assume these are the only ones who seek out organic food—less than 3 per cent of the nation—but 6 million people consume a large amount of food, large enough to be a real market for the organic crops you harvest.

Increasingly, groups are springing up around the country to coordinate the market and help the supply of organic crops get to markets more efficiently. One of the purposes of this publication is to help publicize where organic farmers are so that they can be in contact with food marketing organizations. Even more important is for individual consumers to know about your farm so they can buy from you directly.

Some companies now exist specifically to wholesale or assist the marketing of organic foods. These include:

New Age Natural Foods, Inc.
1326 Ninth Avenue
San Francisco, Calif. 94122
   Fred Rohe

Sun Circle, Max Kozek Produce
651 S. Kohler St.
Los Angeles, Calif. 90021
   Max Kozek

Family Farms Associates of Texas
c/o Malcom Beck
P.O. Box 20318
San Antonio, Texas

Walnut Acres
Penns Creek, Pa. 17862
Paul Keene

Erewhon Trading Corp.
33 Farnsworth St.
Boston, Mass. 02210
Roger Hillyard

Erewhon Trading Corp.
8003 Beverly Boulevard
Los Angeles, Calif. 90048
Paul Hawken

More than 700 retail stores—many of them health stores—which have organic food sections are listed in the *Organic Directory*. If you're interested in developing distribution directly to retail outlets, you may want to have a copy of this list. It's available from Rodale Press, Emmaus, Pa. 18049 for $1.95.

One company that has more than 50 health food stores has offered to make available refrigerator space for organic fruits and vegetables to growers who sign sworn statements. For information, write George McTurk, General Nutrition Corp., 921 Penn Avenue, Pittsburgh, Pa. 15222.

Several companies have begun to advertise that they will provide organic food services. These include: Food & Earth Services, suite 909, 1346 Connecticut Avenue, N.W., Washington, D.C. 20036; Ecological Food Society, 114 East 40th Street, New York, N.Y. 10016.

If you are convinced by now that organic farming can indeed produce a superior product and do it for an indefinite period of time, should you not be able to get a premium for this quality? It is fairly clear that there are an ever-growing number of people looking for just such produce. There is indeed no simple solution to the problem of getting this special producer and discriminating buyer together. However, in many cases, it may be simply a matter of getting to know where they are. And for you to let them know about your farm.

For the livestock producer, quite a distance from urban areas, it may have to involve frozen packages. In fact, this may be the better way to handle meat regardless of the distances involved. Well-operated local meat locker plants may be a good means by which to furnish high quality meat for this ever-growing market. Local cooperatives should get involved in such an enterprise because the processing and marketing of produce should be efficient and have the necessary volume to supply the demand. Actually local fresh meat outlets should not be overlooked because these merchants are not adverse to selling a quality product that their customers want if they can make a profit.

Farmers generally are not in the habit of thinking in terms of marketing. You must begin to become aware of this because much can be gained in *shortening the route from your farm to the consumer*. A little imagination, work and ingenuity will go a long way to find the customer for your organically-grown products.

Be sure to have your farm certified by *Organic Gardening and Farming*. Also be sure to let potential wholesalers and retail outlets of your organic harvests know about your harvests BEFORE you bring in the crop. Work out your arrangements with them as early as possible so you can count on a good market.

The same marketing problems face organic food growers regardless of the area in which they farm, or the length of time they've been in business.

Here are some of the ways it's done:
1. Retail stand situated on or next to farm;
2. Mail deliveries based on food lists sent out by the farm;
3. Delivery route by truck to the home;

4. Wholesale delivery to nearby health-food stores;
5. Wholesale transactions at the farm;
6. Grower cooperatives;
7. Arrangements with specialty food wholesalers.

PART IV

ORGANIC FARM DIRECTORY

CHAPTER 22

Organic Farmer Procedures Questionnaire For 1972

The Organic Gardening and Farming Certification Program is entirely predicated upon your voluntary desire to have the methods and materials which you use and follow verified as organic within the framework of the common definition and consistent with the alternatives which are incorporated in the ideal.

Since each farm, regardless of size, is in fact operated as a separate entity, each farm has unique advantages and problems which are, or should be, best understood by the farmer himself; therefore our certification is not comparative but has actually been designed to verify the procedures developed, or selected, by you. In order to help you verify your own organic way of doing things we need to understand what you are doing and what these procedures can accomplish.

This questionnaire will enable you to give us a qualitative idea of your operation and it will give us tangibles to verify. It is a tool which will be used by Rodale Press representatives and by laboratory technicians to enable us to tell your story more factually, providing consumers with more accurate information and enabling us to supply you with information pertaining to your procedures.

The Organic Gardening and Farming Seal of Certification and inclusion in published listings of Certified Organic Farmers will be awarded on the basis of the information supplied by this questionnaire. As you can see, it is important and essential that you fill out and answer the questions as completely and applicably as possible. Use extra sheets if you wish, and include any additional information or comments which seem useful to you.

NAME_____

ADDRESS_____

_____ZIP_____

PHONE NUMBER (code area)_____

TOTAL ACREAGE_____, TILLABLE_____, COMMITTED_____

Committed organic crops *including* acreage per crop, and estimated harvest volume if more than 2 tons_____

_____

_____

_____

Is committed land owned or leased?_____

Is committed land at another location?_____

If so, please give address or describe location_____

_____

_____

ALL QUESTIONS PERTAIN TO COMMITTED LAND OR ANIMALS

If irrigation water is supplied from an outside source have residue tests been taken?_____

Do we have a copy?_____

Have representative soil samples been taken?_____

Do we have a copy, or copies?_____

Have representative tissue samples been taken?_____

Do we have copies?_____

What is the average humus content?_____

Are there any soil deficiencies?_____Please list if any

_____

_____

_____

If humus content is below 3% what organic materials, and in what quantities per acre, per year, will be used (expectedly) to bring percentage up to 3%?

_____

_____

If humus content is already 3% or more, what materials in what quantities per acre, per year, will be used (expectedly) to maintain humus percentage?

_____

_____

(Tear Out and Mail)

Will additional fertilizer materials be used?_____ If so, please list, (expectedly) giving estimated volume per acre, per year_____

_____

Will other soil conditioners be used? If so, please list

_____

Please list materials, per acre, per year, and the procedures which will be followed to correct any deficiencies of major and minor minerals_____

_____

What weeds are an annual problem? Please list_____

_____

Please describe in detail the weed control procedures which (expectedly) you will follow._____

_____

What insects do you consider to be an annual threat?
Please list_____

_____

What insect control programs will you utilize? Please describe.

_____

Are you annually threatened with fungi diseases? Please list_____

_____

What fungus disease control procedures will you utilize? Please describe.

_____

Are you annually threatened with mildew? Please list_____

_____

(Tear Out and Mail)

71

What mildew control procedures will you utilize? Please describe._____

_____

_____

What soil pests do you consider to be an annual threat? i.e., such as nematodes.
Please list_____

_____

_____

What soil pest control program will you utilize? Please describe._____

_____

_____

_____

Please estimate annual loss from birds, per committed crop. Please list

_____

_____

What bird control program will you utilize, if any? Please describe.

_____

_____

What mosquito control procedure do you utilize? Please describe.

_____

_____

Please list, if any, number and kind of commercial livestock on committed land.

_____

_____

Are any livestock committed for certification? Please list

_____

_____

Will all committed livestock remain upon committed land until sold?_____
Will feeds for committed animals be organically-grown? Please list source of
supplies, grower names, addresses, volume, and kind supplied and how verified.

_____

_____

_____

_____

Source of committed animals, farm?_____ or purchased?_____
If committed animals were purchased what was the average age and weight at
time of purchase? Please list_____

_____

_____

(Tear Out and Mail)

If the age of committed animals was more than one-fourth of the expected maturity at the time of purchase can you verify its history, including preventive shots and feeds? *Please describe in detail* and list names and addresses where animals were purchased._____

_____

_____

Please describe disease and insect control procedures for committed animals.

_____

_____

_____

If committed animals will be committed on a regular basis please include estimated annual production. (If hens have been described include estimated egg production.)_____

_____

_____

Please list equipment and machinery to be used._____

_____

_____

Do you have spraying equipment? If so, please list specific use and the specific materials which you expect to be using relative to organic food production. i.e., dormant oil, water, etc._____

_____

_____

Will you also be using spraying equipment for non-organic production?

_____

What source do you expect to have for organic materials?_____

_____

_____

_____

What source do you expect to have for major and minor minerals?_____

_____

_____

How and where will commodities be stored until sold and removed from your control? Please describe._____

_____

_____

_____

_____

At the present time do you have any agricultural chemicals stored on or in close proximity to committed land, such as bags of manufactured fertilizers, herbicides, and insecticides which are being stored until they can be safely disposed of? If so, please list in detail._____

_____

_____

Do you have any agricultural chemicals stored on or in close proximity of committed land which will be used on non-organic acres? Please describe.

_____

_____

_____

_____

Please describe anticipated crop rotation, if any, to be used on committed land.

_____

_____

_____

As nearly as possible list history of committed land for previous years.

	Last Year	Two Years Ago	Three Years Ago	Four Years Ago	Five Years Ago
Type of fertilizers used including any commercial fertilizers					
Organic matter applied.					
Weed control including, if any, herbicides.					
Insect control including, if any, pesticides.					
Major & Minor minerals.					

On a separate sheet of paper please draw a map locating committed land relative to the nearest community. Please include streets and size of each (if more than one) committed field.

In Western States please return completed questionnaire to:

OGF Certification
Floyd Allen
6203 Toro Creek Road
Atascadero, CA 93422

And in Eastern States return to:

OGF Certification
Rodale Press, Inc.
33 Minor Street
Emmaus, PA 18049

CHAPTER 23

Certified Organic Farmers

The farmers on the following list have been certified through *Organic Gardening and Farming*'s Certification Program (see page 56) as Organic Farmers.

The listing, complete as of the time of publication of this book, includes only California growers. Certifications are continuing, however, and the program is not limited to Californians. Any farmer willing to make the organic commitment and able to meet the standards will be certified.

The listing will be continually revised and up-dated.

Lahai Roi Foundation
Dr. and Mrs. Alan Nittler
3 Lahai Roi Lane
Aptos, CA 95003
(408—688-2257)
Total acreage: 58 acres, 10 acres in avocados
Produce sold: avocados, apples
Sells direct to patients.

Organic Home Nursery
Donn & Rachel Tickner
1688 Pleasant Valley Rd.
Aptos, CA 95003
(408—688-2179)
Total acreage: 2½ acres
Produce sold: Apples, artichokes, and nursery stock. 60% nursery stock started on premise. Other 40% re-potted into organically treated soil.
Sells directly to consumers.

John R. Bigger
Rt. 1, Box 427-G
Arroyo Grande, CA
(805—489-3654)
Total acreage: 42 acres
Produce sold: Squash, corn, Scarlet runner beans, tomatoes, onions, bell peppers.
Sells directly to consumers and to Halcyon Health Food Store.

Baker Bros. Ranch
John C. Baker, Jr.
Box E
Artois, CA
(916—934-4617)
Total acreage: 1502 acres
Committed acreage: 150 acres
Commodity: brown rice
Grown in cooperation with and under the supervision of Wehah Farm for Chico-San Inc.

Mr. & Mrs. Boyd Barker
7065 Sycamore Road
Atascadero, CA 93422
(805—466-6224)
Total acreage: 1½ acres
Produce sold: Corn, green vegetables, berries.
Sells beef and pork not guaranteed to have been fed organically-grown feeds.
Sells directly to consumers.

Joseph and Betty Matthews
7045 Sycamore Road
Atascadero, CA 93422
(805—466-2690)
Total acreage: 1¾ acre
Produce sold: Berries, varieties of fruits and vegetables.
Sells direct to consumers.

New Vineland Farm
Stephen Renquist
8200 Toro Creek Road
Atascadero, CA 93422
Total acreage: 86 acres
20 acres in plums, peaches, apples, corn, pumpkins, sunflower seeds.
Sells directly to consumers.

Santa Lucia Mushroom Farm
Robert W. Bardi
8555 El Corte Road
Atascadero, CA 93422
(805—466-1220)
Total acreage: 38,000 sq. ft.
Produce sold: mushrooms
Sells direct to consumers, retailers, and through Topco Associates and Ziggy's Mushroom and Lime House

John S. Whelen
15300 Morro Road
Atascadero, CA 93422
(805—466-9829)
Total acreage: 80 acres, 11 tillable
Produce sold: apples
Sells directly to consumer. Some sales to retailers.

Ronald T. Epperson
Dry Creek Road
Auburn, CA 95603
(916—878-0236)
Total acreage: 3 acres
Produce sold: turnips, beets, radishes, onions, garlic, lettuce, cabbage, Swiss chard, bell peppers, broccoli, tomatoes, squash, pumpkins, string beans, corn, carrots
Sells direct to consumers.

Bard Valley Citrus Company
R. G. Winder
P.O. Box 4
Bard, CA 92222
(714—572-0063)
Total acreage: 110 acres
Produce sold: oranges, lemons, grapefruit, tangerines, tangelos, medjool dates.
Sells directly to consumers and through distributors, retailers.

Thomson's Tropical Fruits
Paul H. Thomson
Star Route
Bonsall, CA 92003
(714—758-0054)
Total acreage: 10 acres
Produce sold: mangos, topical guavas, cherimoyas, pineapple guavas, avocados, navel oranges, sapotes.
Sells direct to the consumer.

Fraley Lane Organic Orchard
Robert E. Fraley
928 Mission Drive
Camarillo, CA 93010
(805—482-5640)
Total acreage: 1 acre
Produce sold: avocados, Fuerte, Edranal, MacArthur, Nabal, cherimoyas
Sells direct to consumers.

Wolter's Ranch
Russel I. Wolter
Rt. 2, Box 706
Carmel, CA 93921
(408—624-8807)
Total acreage: 60 acres
Produce sold: leaf lettuce, fava beans, sweet corn, apricots
Does not sell directly to consumers. Sells through Frank Capurro, New Age Distributing Co. and Max Kozek Produce Co.

J. H. McKnight Ranch
Donald E. Murphy
P.O. Box 3070
Chico, CA 95926
Total acreage: 1800 acres
Committed acreage: 300 acres prepared for organic production
Commodity: brown rice
Grown in cooperation with and under the supervision of Wehah Farm for Chico-San Inc.

Chet's Organic Farm
Chester H. Lage
12583 East Heather
Clovis, CA 93612
(209—299-7041)
Total acreage: 2½ acres
Produce sold: cherries, apricots, berries, peaches, garlic, tomatoes, squash, watermelon, cantaloupe, grapes, broccoli, cabbage, cauliflower, peas, Brussel sprouts
Sells direct to the consumer.

Good Day Farm
Dennis Broughton and Paulette Koenig
484 Redwood Road
Corralitos, CA 95076
(408—722-1780)
Total acreage: 74 acres
Produce sold: apples
Does not sell directly.

Mr. Harold Warner
P.O. Box 256
Ducor, CA 93218
(209—534-2337)
Total acreage: 40 acres
Produce sold: oranges: Valencia, navel, Terra bella, olives
Sells direct to the consumer.

Keldra Edwards
Rt. 4, Box 472-E
Escondido, CA 92025
(714—746-1353)
Total acreage: 1 acre
Produce sold: oranges
Sells directly to consumers.

Milton H. Parrish
Rt. 4, Box 472 T
Escondido, CA 92025
(714—745-7409)
Total acreage: 4 acres
Produce sold: avocados, tangelos, cherry tomatoes, Mediterranean squash, watermelons
Sells directly to consumers.

Pavone Ranch
Phyllis Pavone
Rt. 4, Box 472-A
Escondido, CA 92025
(714—745-5710)
Total acreage: 5 acres
Produce sold: navel oranges, Valencia oranges, tangerines, tangelos and nuts
Sells direct to the consumer and through retailers.

Dr. Henry S. Stimson
1815 Redwood Street
Escondido, CA 92025
(714—746-3916)
Total acreage: ½ acre
Produce sold: lettuce, carrots, beets, celery, other vegetables
Sells direct to the consumer.

Larry K. Watson
Rt. 2, Box 6039
Escondido, CA 92025
(714—745-8793)
Total acreage: 5 acres
Produce sold: variety of vegetables depending on season and rotation.
Sells directly to consumers and to retailers.

Elk River Organic Garden
Anthony Owen
7532 Elk River Court
Eureka, CA 95501
Total acreage: 6 acres
Produce sold: vegetables, including broccoli, brussel sprouts, and America's strain of Jerusalem artichokes.
Sells to retailers, distributors, New Age Natural Foods, and direct to consumers.

Grovers Organic Grove
D. S. Grover
1758 Prince Street
Fallbrook, CA 92028
(714—728-1979)
Total acreage: 10 acres
Produce sold: avocados, limes, oranges, tangelos, grapefruit
Does not sell direct to consumers. Distributor: Marvelizer Co.

Mr. Sam King
1757 E. Alvarado
Fallbrook, CA 92028
(714—728-2159)
Total acreage: 5 acres
Produce sold: avocados, sapotes, tangerines, oranges, persimmons
Sells direct to consumers.

Grell Pine Hill
John Grell
Grell Hill House
P.O. Box 501
Halcyon, CA 93420
(805—489-2227)
Total acreage: 20 acres
Produce sold: New Zealand Spinach and Christmas trees
Does not sell direct. Distributors: Max Kozek Produce, M&T Produce, Martinis Orchard, Fujishige.

Ronald Waltenspiel
4791 Dry Creek Road
Healdsburg, CA 95448
(707—433-2800)
Total acreage: 50 acres pears, 85 acres prunes.
Distributes only through Timber Crest Farms.

Anderson Farms
John (Jack) Anderson
P.O. Box 486
Knights Landing, CA
(916—662-0895)
Total acreage: 35,000 acres
Committed acreage: 200 acres, may commit more
Produce sold: tomatoes for canning. May also grow broccoli and melons. Tomatoes sold to Sacramento Food, Inc.
NOTE: During growing season Sacramento Foods will supply on request map indicating exact location of committed tomato fields. For other produce inquiries should be directed to farm.

Bonanza Realty
Raymond L. Couch
2071 Riggs Road
Lakeport, CA 95453
(707—263-5386)
Total acreage: 20 acres of walnuts
Produce sold: walnuts, Hungarian paprika
Sells direct to consumer.

Organic Gardens
Emil Sejkora
11230 El Nopol
Lakeside, CA 92040
(213—448-5389)
Total acreage: 1¼ acres
Produce sold: tomatoes, corn, beans, okra, black-eyed peas, chrondu-peas, melons, squash, cantaloupes, pumpkins, winter squash, sunflower seeds
Animals for sale: eggs
Sells direct to consumers.

**Ahlers Organic Date &
   Grape Fruit Garden**
Otto Ahlers
P.O. Box 726
Mecca, CA
Total acreage: 16¾ acres
Produce sold: dates & grapefruit
Sells to retailers & consumers through mail orders. Note: may change hands in 72.

Dave's Rite Fed Meats
David and Jan Hayes
P.O. Box 75
Nelson, CA 95958
(916—342-8864)   (916—343-1977)
Total acreage: 40 acres in permanent pasture. May be increased to 7,000 acres during 1972.
Commodity: Beef. Sold by halves, cut, wrapped and delivered direct to consumers.
NOTE: Not guaranteed to have been fed organically grown feeds until fall; all feeds at least naturally grown. No preventive antibiotics or hormones. Each carcass is analyzed and copy supplied with each order.

John R. Mason
P.O. Box 115
Newberry Springs, CA 92365
(714—257-3231)
Greenhouses
Produce sold: tomatoes
Does not sell direct to consumers
NOTE: Soil being rested and cover-cropped for 1972. No crop expected until January, 1973.

La Malfa Ranch
Anthony La Malfa
Rt. 2, Box 2140
Oroville, CA
(916—882-4224) (Richvale phone)
Total acreage: 630 acres
Committed acreage: 100 acres
Commodities: brown rice
Grown in cooperation with and under the supervision of Wehah Farm for Chico-San Inc.

Del Mar Packing Co.
William B. Witmer
Oxnard & Lompoc, CA
(805—486-3579)
Total acreage: 3000 acres
Committed acreage: 200 acres
Produce sold: lettuce, romaine, cabbage, broccoli, celery, carrots, bell peppers, cucumbers
Produce grown and sold under contract with Max Kozek Produce Co. Note: map locating committed fields supplied upon request.

Olive Spring Farm
Charles E. Snyder
502 Olive Spring Road
Santa Cruz, CA 95060
(408—475-5331)
Total acreage: 12.34 acres
Produce sold: sugar peas, cauliflower, celery, eggplant
Sells direct to the consumer and through New Age Distributing Co.

Rancho Mark West
Barrett A. Johnson
7200 St. Helena Road
Santa Rosa, CA 95404
(707—539-9730)
Total acreage: 400 acres
Produce sold: apples, prunes, cherries
Sells direct to consumer.

Gemini Farm
John Lueschen
Box 604
Solvang, CA 93463
(805—688-4975)
Total acreage: 40 acres
Produce sold: squash, melons, potatoes, horseradish, onions, garlic, peas, beans, corn, artichokes (Jerusalem, globe), lettuce, New Zealand spinach, tomatoes, wine grapes, fruit and nuts, alfalfa and vetch
Animals for sale: eggs
Sells direct to consumer.

Floyd Barsoom
693 South Reed Ave.
Reedley, CA 93654
(408—624-8807)
Total acreage: 20 acres
Produce sold: apricots, Royalty (early)
Does not sell directly. Sells through Ito Packing Co.

Ito Packing Co., Inc.
Mamoru Matsuzaki
Jim Ito
P.O. Box 707
Reedley, CA 93654
(209—638-2531)
Total acreage: 65 acres, 11½ acres farmed organically
Produce sold: 4 acres of Merrill Gem Peaches, 1¾ acres Santa Rosa Plums, 5 acres Laropa-Nubiana plums

Wehah Farm
P.O. Box 216
Richvale, CA 95974
(916—882-4226)
Total acreage: 3,000 acres
Produce sold: rice, oats, vetch
Does not sell direct to the consumer, sells through Chico-San, Inc.

J. B. 2 Farms
Jack DeWit
7904 E. Parkway
Sacramento, CA 95823
(916—422-7536)
Total acreage: 20 acres
Land located 8 miles south of Uba City. Map supplied upon request.
Commodity produced: wheat
Sells to mills, distributors, will sell direct to consumer.

Health Acre
Mildred and Raymond Sponhaltz
17 Kirk Ave.
San Jose, CA 95127
(408—251-3847)
Total acreage: 1 acre
Produce sold: apples, peaches, pears, plums, oranges, potatoes, garlic, onions, green beans, squash, lettuce, corn, boysenberries, strawberries, beets, cabbage, broccoli, chard, spinach, greens, tomatoes, cucumbers
Animals for sale: eggs
Sells direct to the consumer only.

Irwin Ranch
Wesley Messmore
16901 Enchanted Place
Pacific Palisades, CA 90272
(213—454-8067)
Ranch is located at 630 Baylor Street, Lindsay, California.
Total Acreage: 180 acres
Produce sold: Valencia oranges, pomegranates

Rancho San Miguel
Mr. and Mrs. Alan Robertson
R.F.D. Box 176-A
Paso Robles, CA 93446
(805—467-3616)
Total acreage: 130 acres
Produce: fruit, vegetables, grain
Sells directly.

Vanetta Ranch
Carol van Dyke Lundberg
Rt. 1, Box 176
Pleasant Grove, CA
(916—655-3398)
Total acreage: 1000 acres
Committed acreage: 200 acres

Aldren Farms
Alex & Reynald Barsoon
42586 Road 48
Reedley, CA 93654
(209—638-3893)
Total acreage: 74 acres
Committed acreage: 10 acres
Produce sold: peaches
Does not sell directly. Sells through Ito Packing Co.

Tanglewood Farm
Lowell and Freda Curtis
Rt. 1, Box 874-A
Sonora, CA 95370
(209—532-7995)
Total acreage: 8½ acres
Produce sold: apples, pears, peas, lettuce, okra, melons, squash, beans, Swiss chard, tomatoes, onions, horseradish, cucumbers, parsley, cauliflower
Sells direct to consumer by appointment.

Organic Acres Squaw Valley
Robert Gilliland
32042 Tumbleweed Lane
Squaw Valley, Fresno County, CA 93646
(209—332-2526)
Total acreage: 12 acres
Produce sold: summer squash, green beans, garden peas, black-eyed peas, golden beets, cucumbers, sweet corn, carrots, okra, watermelon, cantaloupe, tomatoes, peaches, popcorn, strawberries, green black-eyed peas, apples (Golden Delicious and Crimson Delicious)
Sells direct to consumers.

Ehlen Citrus
Richard W. and Louise M. Ehlen
20976 Road 254
Strathmore, CA 93267
(209—568-2384)
Total acreage: 40 acres
Produce sold: Navel oranges, Valencia oranges, walnuts
Sells direct to consumer and is own distributing co.

Mr. and Mrs. Joseph P. Rubino
374 E. Blewett Road
Tracy-Vernalis, CA 95376
(209—835-7416)
Total acreage: 160 acres committed
Produce sold: apricots, tiltons, blenheims, modesto reds
Animals for sale: chickens and eggs depending on demand
Sells direct to consumers.

Bamboo Garden
Gilbert Rodgers
46 Valley View Road
Watsonville, CA 95076
(408—722-2189)
Total acreage: 35 acres
Produce sold: cucumbers, butternut squash, gold nugget squash, table green squash, tomatoes
Sells direct to the consumers and through New Age Distributing Co.

David and Barbara Klasson
Box 70-K
Ponderosa Way
Whitmore, CA 96096
(916—472-3338)
Total acreage: 75 acres
Produce sold: all varieties of fruit, vegetables
Sells direct to consumers.

The Three Howard Beemans
Howard Dean Beeman, Jr.
Rt. 1, Box 155
Woodland, CA 95695
(916—662-1431)
Total acreage: 7 acres
Produce sold: tomatoes, lettuce, onions, corn
At time of printing, additional committed acreage planned to include tomatoes for canning by Sacramento Foods.
Sells direct to consumer.

Burns Organic Farm
Mr. & Mrs. Edward S. Burns
3308 Woodside Rd.
Woodside, CA 94062
(415—851-1578)
Total acreage: 1 acre
Produce sold: tomatoes, peppers, egg plant, cauliflower, broccoli, squash
Sells directly to consumers, health food stores, and organic restaurant.

Organic Gardening And Farming certifies that all brown rice produced in the years 1971 and 1972 and distributed or processed by Chico-San Inc. and sold under the Chico-San label was/and will be produced by Certified Organic Farmers.

Organic Gardening And Farming certifies that all tomatoes produced in the years 1971 and 1972 and processed by and distributed by Sacramento Foods Inc. and sold under the State Fair label was/and will be produced by Certified Organic Farmers.

CHAPTER 24

Organic Farm Directory

This farm listing is a service without charge to those listed. Most responded to the Organic Farmer Survey (included at the end of the listing), indicating their interest in the organic market. We print the list in hopes of stimulating interaction among all persons, companies and growers interested in organic foods and encouraging more growers to enter the market.

The editors have not investigated each of the individuals or organizations included. As of the time of publication of this list, none of those individuals or organizations included here have been certified as Organic Farmers through Organic Gardening and Farming program. To provide the opportunity for information to reach the public, we must rely on the sincerity of those interested in organic farming.

ALABAMA

M. H. Holmes, Jr.   1000
Childers Bros.   ft
P.O. Box 984
Cullman, Ala. 35055
Livestock: hog, cattle, chickens
Wholesaler

Drexel V. Smith   50
Rt. 2, Box 64
Grand Bay, Ala. 36541
Crops: soybeans
Livestock: cows and calves

Cecil Dyer   40
Harpersville,   pt
Ala. 35078   po
Crops: hay, corn
Livestock: 27 beef cattle
Raise food for my own use.

John R. F. Bond   25
4000 S. Crestview Dr. N.W.   pt
Huntsville, Ala. 35805   po
Crops: family garden crops
Livestock: horses

Gerald L. Wilson   25
Rt. 1, Box 310   pt
Jasper, Ala. 35501
Crops: corn, vegetables, potatoes and fruit
Livestock: 3 hogs
I market my harvest by hand.

Royce L. Earwood   20
Rt. 1, Box 48-B   pt
Mentone, Ala. 35984   ao
Crops: vegetables, some fruit
Livestock: cattle, goats, chickens

Billy Dillard   1½
310 Jones St.   pt
Ozark, Ala. 36360   ao
Crops: tomatoes, cucumbers, sweet potatoes
Quantities: 60, 150 and 40 lbs. weekly as listed
Livestock: fryers and hens
Roadside stand

ALASKA

Charles Fruchan   40
Box 472
Clear, Alaska 99704
Just purchased a small farm and plan to farm it
organically.

ARIZONA

Carryl Baldwin   25
311 Winslow Ave.
Winslow, Ariz. 86047

ARKANSAS

Oren Mason   40
Rt. 1   po
Bald Knob, Ark. 72010
Crop: peas
Livestock: black angus cattle, horses
Roadside stand

Mr. & Mrs. Robert Button   45
Rt. 4, Box 278
Bentonville, Ark. 72712
Interest in organic farming.

Mrs. Vera Barksdale   70
Rt. 1, Box 65   pt
Cave City, Ark. 72521   ao
Crops: okra and tomatoes
Livestock: 30 rabbits and few chickens
Direct to retail store.
Regular commercial channels

Onyx Cave Park   30
Rt. 1, Box 330   pt
Eureka Springs, Ark. 72632   ao
Crops: persimmons, black walnuts, apples,
pears, grapes
Roadside stand and direct to customer

Ferta-Liller Corp.
26 N. College Ave.
Fayetteville, Ark. 72701

George O. Brower 185
Rt. 1 ft
Greenwood, Ark. 72936
Crops: grazing land
Livestock: 23 head cattle

George G. Stiles 60
Rt. 1, Box 6 pt
Leola, Ark. 72084 ao
Livestock: 15 cows and 1 bull

Charles F. Greenstreet 25
Gacross Rt. pt
Melbourne, Ark. 72556 po
Crops: 1 acre grapes, strawberries, vegetables
Livestock: chickens, beef cattle

Mr. Wesley Fenn 146
Rt. 3, Box 292
Ozark, Ark. 72949
Interested in starting an organic farm.

Patsy J. Boyd 200
Rt. 2, Box 136 ft
Paragould, Ark. 72450 ao
Crops: soy beans
Regular commercial channels

O. D. Murphy 160
R. 3, B 236-A pt
Paragould, Ark. 72450 po
Crops: vegetables, grain, hay
Livestock: cattle, hogs

Pat Kuhn ½
Box 23 pt
St. Joe, Ark. 72675 ao
Livestock: 300 chickens, 2 pigs, donkey and 6 ducks

Bill Textor 40
Rt. 1, Box 244 ft
Wilson, Ark. 72395 po
Crops: tomatoes, okra, beef
Sell harvest to private individuals

CALIFORNIA

Lindsey Gunn 50
34 Golfers Lane ft
Bakersfield, Calif. 93308 ao
Crops: navel oranges
Quantities: 10,000 boxes
Regular commercial channels

Sherman S. Hill 250
25749 Community Blvd. ft
Barston, Calif. 92311 po
Crops: alfalfa and cattle
Quantities: 400 tons alfalfa, 100 cows
Livestock: pigs, chickens, ducks, calves
Regular commercial channels

Robert J. Dial 40
Rt. 1, Box 63 ft
Biggs, Calif. 95917 ao
Crops: walnuts, almonds
Quantities: 40 ton walnuts, 2 ton almonds
Regular commercial channels

Paul H. Thomson 4
Star Rt. pt
Bonsall, Calif. 92003 ao
Crops: mangos, tropical guavas, cherimoyas and pineapple guavas.
Quantities: 200, 300, 100, 1,000 lbs. respectively
Regular commercial channels

Hugh J. MacDonald 25
MacDonald Grapefruit Ranch ft
P.O. Box 2106
Calexico, Calif. 92231
Crops: grapefruit
Quantities: 25,000 cartons- 35 lbs.
Wholesaler, direct to retail store and individuals

Robert Fraley ¾
928 Mission Dr. pt
Camarillo, Calif. 93010 ao
Crops: avocados, cherimoyas
Wholesaler or direct to retail store

R. D. Fitzpatrick, Jr. 30
3649 E. International pt
Clovis, Calif. 93612 ao
Crops: figs and wheat
Quantities: 10 tons figs and 6 tons wheat
Regular commercial channels

C. Lage 2½
P.O. Box 82 ft
Clovis, Calif. 93612 ao
Crops: elephant garlic, fruit berries, vegetables, garlic and fruit trees and also grapes
Sell to people who call at home.

Covalda Date Co. 124
Lee J. Anderson ft
P.O. Box 908 ao
Coachella, Calif. 92236
Crops: dates, pecans, citrus
Quantities: 5 to 600,000 lbs. dates
Livestock: hogs
Wholesaler, roadside stand, direct to retail store and mail order.

Mrs. Erin Munoz 25
9128 Haledon Ave.
Downey, Calif. 90240
We are planning to farm in a year from now.

Harold Warner 40
P.O. Box 256 ft
Ducor, Calif. 93218 ao
Crops: oranges
Quantities: 14,000 field boxes
Regular commercial channels

Calvin D. Yandell 1½
619 LaPaloma Rd. pt
El Sobrante, Calif. 94803 ao
Crops: cattle, onions, garlic
Livestock: Angus beef white face, goats, eggs

Pavone Ranch 5
Phyllis Pavone ft
Rt. 4, Box 472 ao
Escondido, Calif. 92025
Crops: citrus (navel, valencia oranges, tangelos, tangerines)
Quantities: 60,000 lbs.
Direct to retail store and mail order.

Henry Stimson ¾
1815 Redwood St. ft
Escondido, Calif. 92025 ao
Crops: lettuce, celery, beets, carrots

Rev. Eli Taplin 1
Rt. 2, Box 2071 pt
Escondido, Calif. 92025 ao
Crops: apples, pears, peaches, walnuts, beets, carrots, lettuce
I sell my harvest privately.
Regular customers buy all my excess fruits and vegetables.

D. S. Grover 10
1758 Prince St. pt
Fallbrook, Calif. 92028
Crops: avocados, oranges, tangelos, limes, lemons
Quantities: 20,000 avocados, 13,000 limes
Livestock: 2 horses, 70 rabbits
Market through a cooperative.

Dan Williams 2½
P.O. Box 36 pt
Fremont, Calif. 94537 ao
Crops: comfrey, corn, beans, oats
Quantities: corn-100 bu. per acre, comfrey-40 tons per acre, oats-3 tons per acre
Livestock: Panama sheep, cattle
Regular commercial channels

Mr. E. Fife 80
131½ N. Berkeley
Fullerton, Calif. 92631
Interested in organic farming.

Ray E. Baker 1½
P.O. Box 183 pt
Gridley, Calif. 95948 po
Crops: pecans, fruit, tomatoes, cucumbers, dill, okra, crenshaw, melons
Quantities: 3,000 lbs.
Direct to retail store and other individuals.

82

Cressle Digby   4
26970 Call Ave.   pt
Hayward, Calif. 94542   ao
Just starting.

Rancher Waltenspiel
  Timber Crest Farms   600
4791 Dry Creek Rd.   ft
Healdsburg, Calif. 95448   ao
Crops: prunes, pears, grapes, apricots
Quantities: several hundred tons
Wholesaler and some mail order for people not
near a health food store.
I market my harvest through health food
distributors.

Edward Sexton   1
9786 Mill Creek Rd.   pt
Healdsburg, Calif. 95448   ao
Crops: prunes-sundried
Sell my harvest through wholesaler and direct
to retail store.

David Petterson   5
Box 543   pt
Kelseyville, Calif. 95451   ao
Crops: vegetables
I give it away to friends.

Edward A. Cross   5
11608 Johnson Lake Rd.   pt
Lakeside, Calif. 92040   ao
Crops: Experimental orchard and garden
Livestock: calves
Direct to retail store

Emil Sejkora   1½
11230 El Nopol   ft
  Organic Gardens   ao
Lakeside, Calif. 92040
Crops: tomatoes, beans, squash, berries
Livestock: cattle, pigs, chickens
Roadside stand

Mrs. Patricia Rudloff
7795 Palm St.
Lemon Grove, Calif. 92045
We are moving into a farming area soon, we
plan organic farming when there.

W. W. Goodman   40
Rt. 2, Box 775   ft
Live Oak, Calif. 95953   ao
Crops: almonds, walnuts
Quantities: 20 tons and 1 ton
Sell to commercial reg. channels and some to
natural food stores.

Charles W. Shepherd   70
P.O. Box 45   ft
Lookout, Calif. 96054   ao
Crops: potatoes, hogs
Livestock: hogs, 100, goats, 30

James A. Doig   1
1872 N. Avenue 56   pt
Los Angeles, Calif. 90042   po
Crops: lettuce, shallots
Livestock: guinea pigs

Louis Farkas   55
319 Lae Dr.
Los Osos, Calif. 93401
Next spring I want to plant a truck garden.

Harold Geiler   5
Star Route   ao
Magalia, Calif. 95954
Crops: carrots, beets, potatoes, corn, tomatoes,
apples, peaches, cherries
Roadside stand, direct to retail store.

Gordon L'Allemand   3
33445 Pacific Coast Hwy.   pt
Malibu, Calif. 90265   ao
Crops: tomatoes, zucchini, squash
Livestock: chickens
Roadside stand and through mail order.

Otto Ahlers   16
P.O. Box 726   ft
Mecca, Calif. 92254   ao
Crops: variety of dates & grapefruit
Quantities: 60,000 lbs.
Roadside stand and direct to retail store and
mail order.

Gene Runnels   25
1214 Grantland Ct.   pt
Modesto, Calif. 95350   ao
Crops: apples, vegetables
Livestock: goats, cattle
Sell to neighbors

Harian Rese   30
Rt. 1, Box 12   ft
Mokelumne Hill, Calif. 95245   ao
Crops: eggs, fruit, nuts, berries
Livestock: chickens, ducks, pheasants, pea
cocks, goats.

Robert MacLean   39
18001 Constitution Ave.   ao
Monte Sereno, Calif. 95030
Crops: french prunes
Quantities: 160 tons (green)
Regular commercial channels

R. E. Dumas   20
197 Mundoro   pt
Morro Bay, Calif. 93442   po
Crops: nuts, fruit, berries

Geo. Damas   96
D & D Natural Food Producers   ft
1384 S. Rice Rd.   ao
Ojai, Calif. 93023
Crops: apples, pears, apricots, peaches, grapes
figs, persimmons, carrots
Quantities: about 40 tons
Wholesaler, roadside stand and direct to retail
store and mail order
Regular commercial channels

Wesley Messmore   80
16901 Enchanted Pl.   ft
Pacific Palisades, Calif. 90272   po
Crops: oranges, pomegranates and olives
Wholesaler
Regular commercial channels

J. P. Broderson   2
723 Cleveland Lane   ft
Petaluma, Calif. 94950   ao
Crops: carrots, tomatoes, cucumbers, cabbage,
peppers, squash
Livestock: rabbits
Regular commercial channels

Laurelwood Acres   45
P.O. Box 577   po
Ripon, Calif. 95366
Crops: roughage to feed our herd of dairy goats.
Our business at present is producing certified
raw goat milk.
Livestock: dairy goats
Would like to convert to all organic produce
raising.

A. B. Stonesifer   40
512 Vine Way   pt
Roseville, Calif. 95678   ao
Crops: irrigated pasture
Livestock: Hereford cattle
Regular commercial channels

E. K. Myers   80
4839 C St.   ft
Sacramento, Calif. 95819   ao
Crops: grapefruits and lemons
Quantities: 8000 boxes

Mrs. Jim Kerdraon   2
1463 Sheridan Rd.   ft
San Bernardino, Calif. 92407   ao
Crops: lettuce, beans, squash

John Briceland   5
26 Genoa Pl.   pt
San Francisco, Calif. 94133   ao
Crops: assorted vegetables
Direct to retail store

Victor Thomson   1500
Rancho Felicia   ft
P.O. Box 748   po
Santa Ynez, Calif. 93460
Crops: hay, grain, cattle, eggs, vegetables
Livestock: beef, dairy cattle, sheep, hens
Regular commercial channels

L. Atietz 94
Stony Brook Ranch ft
22900 Big Basin Way po
Saratoga, Calif. 95070
Crops: prunes, citrus, grapes and vegetables
Quantities: small
Livestock: swine, chickens, beef
Other personal sales
Regular commercial channels

Bob Ross 4
5020 Turner Rd. pt
Sebastopol, Calif. 95472
Crops: fruits and vegetables
Livestock: ducks, few chickens
Sell my surplus direct to retail store.

John Lueschen 42
Box 604 ft
Solvang, Calif. 93463 ao
Crops: vegetables
Livestock: horses, goats, chickens
Direct to retail store

B. L. Curtis 4
Rt. 1, Box 874A pt
Sonora, Calif. 95370 ao
Crops: apples, pears and vegetables
Sell my harvest to neighbors.

Valley Cove Ranch 680
Hugh Gordon ft
P.O. Box 603 ao
Springville, Calif. 93265
Crops: citrus fruits
Quantities: 10,000 bushels
Wholesaler and mail order

Ronald T. Epperson 3
P.O. Box 1195 pt
Tahoe City, Calif. 95730 ao
Crops: diversified vegetables
Livestock: none at present
Roadside stand, health food store and local
restaurant.

Mrs. Joseph P. Rubins
374 E. Blewett Rd.
Tracy, Calif. 95376
Livestock: chickens for our own use
Average 3 tons per acre and hope to attain 10-
15 tons per acre within 5-10 more years.

Mrs. Clarence Perry 30
195-C Hazel Dell Rd. ft
Watsonville, Calif. 95076
Crops: apples
Quantities: 150 tons
Regular commercial channels and others

Mrs. David Klasson
Box 80K, Ponderosa Way pt
Whitmore, Calif. 96096 ao
Crops: truck, orchard
Livestock: milk cow, 3 horses

H. D. Leighty 1
22644 Oxnard St. pt
Woodland Hills, Calif. 91364 ao
Crops: fruits and nuts
Livestock: goats, rabbits, geese and ducks

COLORADO

George D. Luft 800
Arriba, ft
Colo. 80804 po
Crops: wheat
Quantities: 4,000 bushels
Livestock: 70 head of cattle
Regular commercial channels

John N. Strahan 80
Rt. 1, Box 85 ft
Boone, Colo. 81025 ao
Crops: fesh vegetables, melons, beef
Livestock: chickens, pork, beef
Direct to retail store and give some away

Albert L. French 40
800 Linden Ave. pt
Boulder, Colo. 80302 ao
Crop: alfalfa
Quantities: 4 tons per acre
Livestock: chickens, sheep and horses
Sell harvest to neighbor who feeds it to his dairy
herd or retails it if there is a surplus.

Don Roberts 400
Rt. 1, Box 108 ft
Del Norte, Colo. 81132 po
Crops: hay & barley
Quantities: 500 tons hay, 3000 bu. barley
Livestock: 400 sheep, 80 cows
Regular commercial channels

Dr. Lawrence Ensor 5
3270 Meade pt
Denver, Colo. 80211 ao
Crops: fruits and vegetables
Livestock: 20 head angus
Direct to retail store and I own the store.

Paul Bethke 6
Rt. 3, Box 540 pt
Ft. Collins, Colo. 80521 ao
Crops: sweet corn, and all garden vegetables
Quantities: 500 doz. corn
Livestock: lamb, beef, organic eggs
Roadside stand

Mr. & Mrs. Dale Dossey 4
14141 W. 72nd Ave.
Golden, Colo. 80401
Interested in Organic Gardening.

Robert Lawless 350
884-24½ Rd. pt
Grand Junction, Colo. 81501 ao
Crops: wheat and grain
I sell my harvest direct.

John D. Schmahl 80
2301 River Rd. pt
Grand Junction, Colo. 81501 po
Crops: wheat and corn
Quantities: 60 tons wheat, 90 tons corn
Regular commercial channels.

Mrs. Martin Schobert 100
Box 17A, Star Rt. pt
Hudson, Colo. 80642 po
Crop: alfalfa
Livestock: 16 holstein, 47 angus

Norman Huffaker 12
Rt. 4, Box 123A pt
Longmont, Colo. 80501 ao
Crops: alfalfa and chickens, eggs
Quantities: 30 tons alfalfa
Livestock: chickens, geese, ducks
Direct to retail store and off the farm.

Mrs. Ted Ryden 250
Rt. 1, Box 49 ft
New Castle, Colo. 81647 po
Crops: hay, alfalfa, cattle and grain
Quantities: 500 tons hay and 1000 bu. grain
Livestock: 250 head Hereford cattle, 15 tur-
keys, 100 chickens and 6 horses
Market through regular channels

Michael Cummings 40
Sunshine Mountain ft
Star Rt. ao
Paonia, Colo. 81428
Crops: cherries, pears, peaches, plums, apples
Quantities: 4 tons, 10 tons, 1000 lbs., 2 tons
20 tons
Roadside stand
Regular commercial channels.

Wildwood Ranch 50
P.O. Box 25 ft
Paonia, Colo. 81428 ao
Crops: oats, wheat, apples
Quantities: 400-1200 bu. apples, 20 tons each
wheat, oats
Livestock: hogs, cows, chickens
Sell my harvest direct to consumer.
Regular commercial channels

Alice M. Irvin
Box 197
Penrose, Colo. 81240
Livestock: 25 to 30 horses

Paul T. Bechtol, Jr.   43
Rt. 1, Box 4A   pt
Peyton, Colo. 80831
Would like to start farming soon.

Paul H. Kerschner   1100
Rt. 3   ft
Sterling, Colo. 80751   ao
Crops: wheat, milo, millet, barley
Quantities: 6000 bu. wheat, 3000 milo, 1000
millet, 400 wheat
Livestock: stock cattle
Regular commercial channels

Robert Wade
Walden,   ft
Colo. 80480   ao
Crops: potatoes
Livestock: chickens & strawberries

Geba B. Hannon   160
R. 1, Box 63   ft
Wiggins, Colo. 80694   po
Crops: sugar beets, corn, pinto beans, alfalfa,
carrots, cucumbers
Quantities: 30 acres beets, 35 acres corn, 20
acres beans, 50 acres alfalfa, 5 acres carrots, 10
acres cucumbers
Direct to retail store and direct to consumer.
Regular commercial channels

CONNECTICUT

Ernest A. Mannel   5½
Beacon Rd.   ft
Bethany, Conn. 06525   ao
Crops: strawberries and vegetables
Roadside stand

Mark Wahlberg   10
Albany Tpke.   pt
Canton, Conn. 06019   ao
Crops: potatoes, corn
Quantities: 1 acre of each
Livestock: beef cattle, hogs
Direct to retail store.

Perry McMaster   8½
Isaiah Smith Lane   pt
E. Morris, Conn. 06763   ao
Crops: tomatoes, cucs, corn, caluiflower, beans

F. Abston   3
P.O. Box 227   pt
Gaylordsville, Conn. 06755   ao
Roadside stand and direct to retail store.

Pulsa   1
282 Riggs St.   pt
Oxford, Conn. 06483   ao
Crops: beans, kale, carrots, broccoli, peas
Quantities: 100 lb. per week
Roadside stand and direct to retail store.

Malabar Farm   42
Glory Rd.   ft
Weston, Conn. 06880   ao
Crops: winter squash
Regular commercial channels and food asso-
ciations.

DELAWARE

Mr. & Mrs. Zed D. Goodson 100
Rt. 1, Box 306
Willow Grove
Camden, Del. 19934
Expect to be building an organic truck farm
and to have a roadside stand.

FLORIDA

Adolf Woernle, Jr.   15
311 60th St. N.W.   ft
Bradenton, Fla. 33505   ao
Crops: all types of citrus
Quantities: 3000 bu.
Sell direct mail.

Floyd K. Walker   1
Rt. 1, Box 152   pt
Brooksville, Fla. 33512   po
Crops: frying rabbits and goat milk
Livestock: rabbits, goats and chickens

Robert M. Jernigan   35
P.O. Box 1   pt
Ft. Pierce, Fla. 33450   ao
Crops: citrus—oranges and grapefruit
Quantities: 2,000 boxes

Donald & Dorothy Bennett
RFD 1, Box 151
Fountain, Fla. 32438
Livestock: goats, fowls, hogs, rabbits
Just relocated to our own land of 60 acres.

Herman W. Woods   5
Rt. 1, Box 21   pt
Fountain, Fla. 32438   po
Crops: pecans, English walnuts, figs
Quantities: 1200 lbs.
Regular commercial channels

Ken Bynum   100
Rt. 1, Box 51   ft
Jay, Fla. 32565   po
Crops: soybeans, wheat, oat
Quantities: 160 tons

Wallace Johnson, Jr.   300
P.O. Box 63   pt
Ocala, Fla. 32670   po
Crops: hay and pasture

Joseph Cepuran   5
Rt. 1, Box 114D   pt
Sanford, Fla. 32771   ao
Crops: most vegetables, strawberries
Direct to retail store.

F. W. Carraway, Jr.   150
P.O. Box 29   pt
Tallahassee, Fla. 32302   po
Crops: corn, game food
Livestock: 42 cows, 86 steers

Richard H. Sloan   65
Rt. 7, Box 1259   ft
Tallahassee, Fla. 32303   po
Crops: vegetables
Wholesaler and roadside stand.

GEORGIA

Claude Lee   27
Rt. 3, Box 9   pt
College Park, Ga. 30337
Crops: trees, want to start farming at least 15
acres of organic vegetables.

Wilbur O. Clough   10
2239 Pine Wood Dr.   pt
Decatur, Ga. 30032   ao
Crops: all kinds of vegetables
Livestock: 2 cows, 5 goats, 1 hog and chickens
Roadside stand

W. C. Crowley, Jr.   15
3402 Jackson Dr.   pt
Decatur, Ga. 30032   po
Crops: corn, field peas and garden
Quantities: 300 bu. corn and 100 bu. peas
Livestock: 2 cows, 3 pigs, 60 hens and 1 mule

Charles R. Short   1
108 Scott Ave.   pt
Griffin, Ga. 30223   ao
Crops: tomatoes, beans, squash, corn, cucumber
I give my harvest away and am building a
market.

Randall Ransom   100
Triple Creek Farm   ft
Hemp, Ga. 30515   ao
Crops: grains and seed corn
Livestock: goats, chickens, bees

Marlin Glin Clark   100
Rt. 1, Box 209   pt
Rocky Face, Ga. 30740   po
Crops: hay & cattle
Livestock: cattle
Regular commercial channels

Miles F. White   30
Rt. 5, Box 72   ft
Rome, Ga. 30161   po
Crops: hay
Livestock: goats, cows and horses

Pine Hills Herb Farm   2
P.O. Box 307   ft
Roswell, Ga. 30075   ao
Crops: herb plants and subsistance garden for
two.
Livestock: 2 goats, 1 steer, 12 chickens
Regular commercial channels

H. Van Der Borden   14
Box 686   pt
Roswell, Ga. 30075   ao
Crops: vegetables, strawberries
Livestock: chickens
Direct to retail store and individuals.

Edward D. Colston   6
Rt. 1   ft
Taylorsville, Ga. 30178   ao
Crops: tomatoes, beans, potatoes and turnips
Quantities: 5 to 7 tons tomatoes, 75-100 bu.
beans, 40 to 50 bushels of potatoes and turnips
Regular commercial channels

HAWAII

J. O. Catoagan   29
1937-A St. Route   ft
Kaneohe, Ha. 96744   po
Crops: bananas, papayas, plants, flowers
Livestock: pigs

ILLINOIS

Edward Johnson   160
Benson,   ft
Ill. 61516
Crops: corn, oats, beans, hay
Livestock: 22 head cattle
Regular commercial channels

Meinert Gronewald   70
Rt. 2   pt
Carthage, Ill. 62321   po
Crops: corn, soybeans, wheat, oats
Quantities: 1000 bu. corn, 400 bu. soybeans, 200
bu. wheat and 300 bu. oats
Livestock: cattle and poultry
Regular commercial channels.

Ruth Berk
7052 N. Greenview   ao
Chicago, Ill. 60626
Crops: tomatoes, beans, melons
Livestock: beef, chickens
160 acres only a small area is under cultivation.

Dennis L. Cox   40
3136 W. 54th Pl.
Chicago, Ill. 60632
Would like to convert to full organic methods.

James A. Handley   10
1825 N. Lincoln Plaza   ao
Chicago, Ill. 60614
Crops: fruit and vegetable variety
Quantities: sufficient to supply 4 retail stores

Arthur Kulosa, Jr.   39
6026 N. Oconto Ave.   pt
Chicago, Ill. 60631   po
Crops: vegetables, cattle, watercress, berries,
nuts
Quantities: small
Livestock: cattle
Mostly self used, but if it would pay we could
sell to the organic market.

Walter J. Sluzas   30
12237 S. Yale Ave.
Chicago, Ill. 60628

Emery Schwengel   120
Rt. 2   ft
Dieterich, Ill. 62424   ao
Crops: corn, beans
Livestock: 30 cattle, 100-150 hogs

Edward W. Tucker   40
R.R. 5   pt
East Peoria, Ill. 61611   ao
Crops: soybeans, cattle, vegetables
Quantities: 20 acres beans, 10 acres hay
Livestock: 10 head beef
Regular commercial channels

Mrs. W. T. Quentin   70
Glendale Organic Farm   ft
Box 318, R.R. 2   ao
Effingham, Ill. 62401
Crops: corn, wheat, oats, hay, rye and veg.
Livestock: 2 cows, several chickens, 3 angus
cows, 2 angus heifers, 1 angus Jersey heifer
Regular commercial channels, surplus veg. and
beef. I sell direct to buyers through contact of
your shoppers guide.

Edward Wagoner   30
1217 W. Stephenson   ft
Freeport, Ill. 61032
Have a small garden and plan to farm all-or-
ganic soon.
Livestock: beef cattle, chickens and some sheep

Thaddeus J. Podolak   25
45 Park Ct.   pt
Glen Eilyn, Ill. 60137   ao
Crops: fruit trees and vegetables
I will sell my harvest to wholesaler and road-
side stand.

Ezra Boyer   420
R.R.   ft
Green Valley, Ill. 61534   ao
Crops: corn, soybeans, oats
Livestock: 400 hogs, 50 cattle
Regular commercial channels

Joe Rench   15
R.R. 2   pt
Greenville, Ill. 62246   po
Crops: strawberries, tomato, and other vege-
tables
Quantities: 3 acres strawberries
Livestock: 2 ponies, 2 beef steers
Regular commercial channels

Francis Rinderer   500
503 Prairie   ft
Greenville, Ill. 62246   po
Livestock: cattle
Regular commercial channels

Curtis Stojan   74
R.R. 2   pt
Hampshire, Ill. 60140   po
Crops: corn, sweet corn, pickles, melons,
strawberries, raspberries
Quantities: 6000 bu. corn, 3 acres raspberries
& strawberries, 2 acres pickles and melons
Livestock: chickens, geese, ducks, beef cows
Roadside stand

Ralph Walters   42
Industry,   ft
Ill. 61440   po
Crops: corn, wheat, soybeans, clover
Quantities: corn 1000 bu., wheat 200 bu. and
soybeans 200 bu.
Livestock: 9 calves a year
Regular commercial channels

Glenn R. Urban   120
5269 Meyer Dr.   pt
Lisle, Ill. 60532   po
(Farm in Marshfield, WS)
Crops: corn, oats, alfalfa, beans
Livestock: dairy cows

Ernest Halbleib 170
Halbleib Orchards ft
R.R., P.O. Box 42 ao
McNabb, Ill. 61335
Crops: corn, wheat, rye, soybeans
Quantities: 20 acres corn, 20 acres wheat, 18
acres rye, 60 acres soybeans
Livestock: Holstein cattle, goats
Wholesaler, direct to retail store and store on
farm.

William Edmund Frazer 32
RFD 1 pt
Neponset, Ill. 61345 ao
Crops: fruit and vegetables, honey and beef
Livestock: 14 black Angus
I sell to friends and neighbors.

Fr. Ambrose German, O.F.M.
St. Joseph's Franciscan Seminary
3313 Midwest Rd.
P.O. Box 449
Oak Brook, Ill. 60521
Crops: tomatoes, cucumbers
Livestock: dairy herd and good number of hogs

Harold N. Simpson 80
144 W. Harrison pt
Oak Park, Ill. 60304 ao
Crops: grain
Livestock: pigs

Harry Wagler 54
119 S. Market St. pt
Paxton, Ill 60957 po
Crops: corn, oats, wheat, soybeans
Regular commercial channels

Horace A. Mollet 100
Pocahontas, ft
Ill. 62275 po
Crops: corn, soybeans, vegetables
I sell my harvest to wholesaler.
Market through regular commercial channels

Paul Hess 120
Prairie Du Rocher, ft
Ill. 62277 ao
Crops: corn fed to livestock
Livestock: 150 hogs, 15 to 20 cattle
Regular commercial channels

M. & L. Fruit Farm 25
Rt. 1, Box 115 pt
Prophetstown, Ill. 61277 po
Crops: fruit and vegetables
Livestock: sheep

Hilmer Swanson 60
9 McKee Dr.
Quincy, Ill. 62301
Will be moving in March and hope to start
farming. Hope to establish a small health food
store at the farm to sell food supplements along
with the fresh food.

Max M. Dally 65
St. Augustine, ft
Ill. 61474 po
Crops: soybeans and corn
Quantities: corn—80 bu. to 100 bu. per acre,
beans—25 to 30 bu. per acre
Livestock: 3 cattle
Regular commercial channels

Ronald Russell 3
1503 Maryland Ave. pt
Springfield, Ill. 62702 ao
Crops: all types of vegetables

Stan Durin 220
R.R. 1 pt
Steward, Ill. 60553 po
Crops: corn, soybeans, wheat, oats
Quantities: 125 bu. corn, 75 bu. soybeans, 15
bu. wheat, 5 bu. oats
Livestock: hogs, beef cattle
Regular commercial channels.

Monroe Werling 75
R.R. 2 po
Waterloo, Ill. 62298
Crops: wheat, corn
Regular commercial channels

J. L. Reimer 35
R.R. 3 pt
W. Frankfort, Ill. 62896 ao
Crops: vegetables
Livestock: goats, rabbits, quail

Arthur Michels 300
Rt. 1 ft
Wheeler, Ill. 62479 po
Crops: soybeans and corn, wheat
Livestock: beef cows
Regular commercial channels

Wagoners Organic Acres 12
Box 116 pt
Williamsfield, Ill. 61489 ao
Crops: green beans, sweet corn, tomatoes, pep-
pers and garden vegetables
Livestock: cattle, hogs, poultry, ducks
Roadside stand at farm.

IDAHO

Robert Trecker 30
Rt. 1, Box 253 pt
Cataldo, Id. 83810 ao
Crops: garden vegetables and hay
Livestock: rabbits, chickens, beef

INDIANA

Fosters Gardens 7½
102 E. Cleveland St. pt
Alexandria, Ind. 46001 po
Crops: soybeans, corn, chrysanthemums
Quantities: 226 bu. beans, 1000 mum plants
Regular commercial channels

Morton Deeter 80
R. 3, Box 52 ft
Anderson, Ind. 46011 po
Crops: corn, hay, sweet corn, beans
Livestock: 100 hogs
Roadside stand

Clinton Goins 179
Rt. 5, Box 22 ft
Anderson, Ind. 46011 po
Crops: soybeans, rye, oats, corn, wheat, hay
and beef
Quantities: 26000 bu. beans, corn every other
year, 20 acres of oats, rye and wheat
Livestock: around 40 head half beef half
springer holstein heifers.
Regular commercial channels and sell to pri-
vate customers.

John McMahan 21
Box 111 ft
Clifford, Ind. 47226 ao
Crops: beef, poultry, strawberries, rhubarb,
asparagus, peas
Livestock: cattle, chickens, hogs, goats
Roadside stand at the farm.

Jesse Hartzog 20
R. R. 3 pt
Columbia City, Ind. 46725 po
Crops: corn, wheat, oats, rye, barley
Regular commercial channels

Dale D. Treesh 200
Rt. 1 ft
Corunna, Ind. 46730
Crops: corn, wheat, oats
Quantities: 5000 bu. corn, 1000 bu. wheat, 2000
bu. oats
Livestock: 150 head sheep
Regular commercial channels

Raymond Martin 40
R.R. 1, Box 36 pt
Covington, Ind. 47932 ao
Crops: corn, oats, alfalfa, clover, hay
Quantities: 800 bu. corn, 300 bu. oats, 10 tons
hay
Livestock: hogs, sheep
Regular commercial channels

Bill Price ½
2340 Morton Ave. pt
Elkhart, Ind. 46514 po
Crops: vegetables
Livestock: ducks, chickens, geese, guinea
Home delivery

Doyle Robinson 30
Rt. 2 pt
Fremont, Ind. 46737 po
Livestock: horse, chickens

Chester Haffron 15
R. R. 1 ft
Hazleton, Ind. 47540 ao
Crops: tomatoes, melons, berries
Livestock: goats and chickens
Regular commercial channels

W. Marvin Lundy 2½
Lone Pine Farm pt
Inglefield, Ind. 47618 ao
Crops: lettuce, corn, beans, potatoes, celery,
onions
Livestock: beef
Local pick-up and direct by mail order

O. C. West 5
Rt. 1, Box 68186 ao
Lakeville, Ind. 46536
Crops: alfalfa, potatoes, carrots, onions
Quantities: 5 bu. carrots, 20 bu. potatoes, 600
lb. onions, 200 bales alfalfa
Most of my sales are local.

Michael L. West 10
603 E. Walnut St. pt
Lebanon, Ind. 46052 po
Crops: tomatoes, melons, strawberries
Livestock: rabbits and chickens
Roadside stand

Thomas C. Bennett 20
R.R. 5 pt
Logansport, Ind. 46947 po
Crops: hay
Quantities: 1000 bales
Livestock: 20 calves, 6 hogs
Regular commercial channels

Lee R. Cory 80
Rt. 1 pt
Milford, Ind. 46542 po
Crops: melons, corn, squash, sweet potatoes,
tomatoes
Livestock: 20,000 ducks annually
Direct to retail store
Regular commercial channels

Lone Organic Farm 62
R. 1 ft
Millersburg, Ind. 46543 ao
Crops: hay, pasture, garden
Livestock: a few cows, young cattle
Sell direct to customers.

Michael B. Nall 30
R.R. 2, Box 286
Mitchell, Ind. 47446
Livestock: cattle

Mr. & Mrs. Byron E. Green 75
R.R. 1, Greendell Farm pt
Mooresville, Ind. 46158 ao
Crops: hay, pasture, strawberries, honey, corn
Livestock: 110 ewes and a few geese
Wholesaler

Veda Brenneman 40
R.R. 6 ft
New Castle, Ind. 47362 ao
Crops: corn and soybeans
Quantities: 80 bu. corn, 24 bu. beans
Livestock: 2 goats
Regular commercial channels

Stanley Ness 250
61651 Crumstown Tr. pt
N. Liberty, Ind. 46554 ao
Crops: corn, wheat, soybeans, oats
Quantities: 6000 corn, wheat 500, soybeans
2000, oats 1000
Livestock: 50 head beef cows
Regular commercial channels

Thomas Werner 50
R. 3 ft
Osgood, Ind. 47037 ao
Crops: corn, wheat, hay, vegetables
Livestock: beef cattle, hogs, chickens
Direct to retail store
Regular commercial channels

De Motte-Evans Farms 250
Otwell, ft
Ind. 47564 ao
Crops: corn, wheat and soybeans
Quantities: 5,000 corn, wheat 1,000, soybeans
2,500
Livestock: turkeys and chicken fryers
Wholesaler and poultry available at farm.
Regular commercial channels

Jack Coke 35
R. R. 1 ft
Paoli, Ind. 47454 po
Crops: corn
Quantities: 150 bu. per acre
Livestock: 5 cows, 12 chickens, 18 hogs

Max E. Wilson 80
R.R. 3 pt
Paoli, Ind. 47454 po
Crops: honey and dairy products
Livestock: 12 head beef

Ralph D. Foulk 70
R.R. 3, Box 712 pt
Ray, Ind. 46737 po
Crops: corn, alfalfa, oats, sweet clover
Quantities: 60 to 100 bu. per acre
Livestock: 70 chickens
Regular commercial channels

Robert Earehart 20
R.R. 1, Box 236 pt
Redkey, Ind. 47373 po
Crops: corn, tomatoes, beets, peas, beans
Livestock: 3 horses
Roadside stand

James L. Helton 40
R. 24, Box 76 pt
Terre Haute, Ind. 47802 ao
Crops: beef and honey
Quantities: 4-6 beef, 300 lb. honey
Wholesaler, roadside stand

Rex Pague 30
Rt. 22, Box 271
Terre Haute, Ind. 47802
Crops: soybeans
Livestock: Ancos chicken & 3 mule-footed hogs

Robert W. Hager 32
R.R. 2, Box 120 pt
Valparaiso, Ind. 46383 po
Crops: corn, soybeans, vegetables, fruit
Quantities: 1000 bu. corn
Livestock: Angus cattle, hogs, chickens, ducks,
geese
Regular commercial channels

Walter M. Thompson 53
2018 Washington Ave. ao
Vincennes, Inc. 47591
Crops: grass
No crops have been removed from this land
for approx. fifteen years.

Ronald Rhoades 40
Rt. 1 ft
Winchester, Ind. 47394 po
Crops: corn
Quantities: 80 bu. per acre
Livestock: hogs, cattle, chickens, rabbits, 1
dairy cow
I feed my harvest to hogs.

Mr. & Mrs. John Wallen 18
R. 1, Box 304A pt
Yorktown, Ind. 47396 ao
Crops: corn and vegetables
Livestock: hogs
I sell my harvest through newspaper adver-
tisement.

IOWA

Mahlon Woodley 190
Rt. 2 ft
Allison, Iowa 50602 ao
Crops: corn, beans, oats, rye
Livestock: sheep, hogs, dairy
Sell harvest to wholesaler and direct to customer.
Regular commercial channels

Royal Dean Alexander 40
3718 Ross Rd. pt
Ames, Iowa 50010

Richard L. Thompson 300
Rt. 2, Box 164 ft
Boone, Iowa 50036 ao
Crops: corn, soybeans, hay, oats, grains
Livestock: cattle, hogs
Regular commercial channels

Ernest Witzel 175
Castana, ft
Iowa 51010 ao
Crops: alfalfa, hay, corn, oats
Livestock: milk cows, 100 head hogs year
Regular commercial channels

Wm. Schwab 63
Columbus Junction, pt
Iowa 52738 po
Crops: oats, corn, soybeans
Quantities: 400 bu. beans, 400 corn, 400 oats
Livestock: 10 cows
Regular commercial channels

Mrs. Clarence Ryon 55
R.R. 1 pt
Dayton, Iowa 50530 po
Crops: hay, vegetables for family
Livestock: goats and horses

Dale R. Reiser 349
Dorchester, ft
Iowa 52140
Crops: corn, hay, oats
Quantities: 18,000 bu. corn, 20,000 tons hay, 1,200 bu. oats
Livestock: beef cows, dairy, fat cattle, hogs
Regular commercial channels

Peter G. Henkels 2
1610 Kane St. pt
Dubuque, Iowa 52001 ao
Crops: melons and potatoes
Quantities: 1500 watermelons, 1200 muskmelons and a ton potatoes
Roadside stand and direct to consumer.

Glen Fuhrman 70
Dundee, ft
Iowa 52038 po
Crops: corn, oats, barley, wheat
Livestock: cattle and hogs
Regular commercial channels

Roger Sheldon 160
Rt. 1 ft
Dundee, Iowa 52038 ao
Crops: corn, oats, soybeans, hay
Quantities: 500 bu. corn, 1800 bu. oats, 50 tons hay
Livestock: beef cattle, sheep

Phillip Trierweiler 340
Earling, ft
Iowa 51530 po
Crops: corn, oats, soybeans
Quantities: 10,000 bu. corn, 3000 bu. oats, 2000 bu. soybeans
Livestock: cattle, hogs, chickens
Regular commercial channels

LaVerne Penning 152
Geneva, ft
Iowa 50633
Crops: corn, soybeans, oats and hay
Quantities: 6500 bu. corn, 1200 beans, 1000 oats
Livestock: hogs and beef cattle
Regular commercial channels

Merlyn Steer 153
Greene, ft
Iowa 50636 ao
Crops: corn, oat, wheat
Livestock: hogs, cattle

Clarence Van Sant 120
Rt. 2 ft
Grinnell, Iowa 50112 ao
Crops: corn, hay, soybeans, oats
Quantities: 2500 bu. corn, 80 tons hay, 1200 bu. soybean, 1200 bu. oats
Livestock: 27 charolais cows, 1 bull, 300 head of hogs a year

Fred Trafelet 90
R.R. 2 ft
Hawkeye, Iowa 52147 po
Crops: corn, oats and hay
Quantities: corn—2000 bu., oats—1000 bu., hay—75 tons
Livestock: beef, hogs and poultry
Sell to general buyers as a produce station, hog buying station and livestock auction.
Market to some buyers that come and buy.

David Weber 145
R. 1 ft
Hedrick, Iowa 52563 po
Crops: corn, soybeans, oats and hay
Livestock: cattle, hogs
Regular commercial channels

Richard Veningo 100
R.R. 2 ft
Hull, Iowa 51239 ao
Crops: corn, oats, alfalfa
Livestock: heifers
Regular commercial channels

Mrs. Robert W. Weeber 20
R.R. 1, Box 218 pt
Iowa City, Iowa 52240 ao
Crops: varied

Roger Boldt 45
RR #1, Box 86 pt
Kalona, Iowa 52247 ao
Crops: vegetables

Clarence Langstroat 265
Larchwood, ft
Iowa 51241 ao
Crops: corn, hay, oats
Livestock: cattle, dairy cows, hogs, chickens, and geese
We use most of our feed for our livestock.

Mr. & Mrs. Harlan Abbas 118
Rt. 1 ft
Latimer, Iowa 50452 po
Crops: corn, peas, apples, small veg. garden
Livestock: 50 sheep, 30 hogs, 8000 chickens, 50 ducks, 25 guineas

Lenus D. Meanne 170
Rt. 1 ft
Luana, Iowa 52156 po
Crops: corn, hay, oats
Livestock: milk cows, head hogs

G. O. Towne 14
Manson, pt
Iowa 50563 po
Crops: corn, soybean, alfalfa
Regular commercial channels

Donald DeKlotz 240
R.R. 1 ft
Newhall, Iowa 52315 po
Crops: corn, oats, hay pasture, soybeans
Quantities: 10,000 bu. corn, 3500 bu. oats, 5,000 bales hay, 400 bu. beans
Livestock: cattle, hogs and chickens
Regular commercial channels

Roger J. Hunt 27
RR 1 pt
Nora Springs, Iowa 50458 po
Crops: beans, corn, hay
Quantities: 290 bu. beans, 800 bu. corn

Merlin A. Faas 195
North English, ft
Iowa 52316 ao
Crops: corn, oats, hay & pasture
Livestock: 52 Reg. Angus cattle, hogs
Regular commercial channels

George Nelson 300
R.R. 4 pt
Osceola, Iowa 50213 po
Crops: corn, beans, oats, hay, garden products
Livestock: cattle and hogs
Regular commercial channels

Mr. & Mrs. Donald Coe 40
Box 100 pt
Rock Falls, Iowa 50467 ao
Crops: corn, oats and beans
Quantities: 10 acres each
Sell my harvest through grain elevator near by.

Wilfred Wagenaar 160
Rt. 2, Box 78 ft
Sheldon, Iowa 51201 ao
Crops: corn, soybeans, oats, hay
Quantities: 4000 corn, 1500 beans, 1000 oats,
2000-3000 bales hay
Livestock: 12 cows and calves, 15 bulls, 100-
200 hogs
Regular commercial channels

Donald R. Bedill 115
Springville, ft
Iowa 52336 ao
Crops: corn, oats, hay, soybeans
Quantities: 33 acres, 20 acres, 20 acres, 15 acres
Livestock: beef cattle and hog
Regular commercial channels

Thomas Mansheim 45
R.R. 1 pt
West Point, Iowa 52656 ao
Crops: hay & beef
Quantities: 2000 bales of hay
Livestock: 19 head cattle
Regular commercial channels

Mrs. Ana Anderson 55
West Union, ft
Iowa 52175 ao
Crops: oats, corn, hay
Quantities: 600 bushels corn, 300 or more oats,
2000 to 3000 bales of hay
Livestock: 12 milk cows, 12 growing heifers
and 4 butcher Holstein calves
Sell milk to open market and heifers to farmers
around the country.

KANSAS

Keith Hurst 30
R.R. 2 pt
Arkansas City, Kansas 67005 ao
Crops: wheat, alfalfa and tomatoes
Livestock: cattle
I sell to friends and relatives

Mrs. Jon Yenni 1000
Glasco, ft
Kansas 67445 po
Crops: wheat, milo, alfalfa, soybeans
Regular commercial channels

John R. Vogelsberg 335
Home, ao
Kansas 66438 ft
Crops: alfalfa, corn, soybeans, and wheat
Quantities: corn—3000 bu., soybeans—1000
bu., wheat—1500 bu.
Livestock: 30 cattle, 75 hogs and 150 hens
Regular commercial channels

Abe R. Thiessen 130
Inman, pt
Kansas 67546 ao
Crops: wheat, milo, alfalfa
Quantities: 1200 bu. wheat, 1500 milo and 20
ton alfalfa
Livestock: cows, calves and hogs
Regular commercial channels

Alvine Schmedemann 100
R. 1 ft
Junction City, Kansas 66441 ao
Crops: corn, wheat, apples, alfalfa
Quantities: 1,000 bu. wheat, 200 bu. apples,
200 ton alfalfa
Livestock: beef cattle
Regular commercial channels

Wesley Wilson 200
Box 481 pt
LaCrosse, Kansas 67548 ao
Crops: wheat
Quantities: 5000 bu.
Regular commercial channels

Jack Dwerlkotte 750
Marysville, ft
Kansas 66508
Crops: alfalfa, wheat, corn
Quantity: 200 acres wheat, 20 acres corn

Marvin Ratzlaff 2½
RFD 1 ao
St. George, Kansas 66535
Crops: garden vegetables
Plan to supply a local health food store.

Tomlene Smith 40
RFD 1 ao
Tongamoxie, Kansas 66886 po
Crops: milo
Quantities: 14 acres
Livestock: 4 cows and 4 goats

B. J. Heeney, Jr. 300
Rt. 2
Topeka, Kansas 66608
Crops: cotton, alfalfa
Livestock: 40 head Angus-Hereford
Regular commercial channels

Shelby Leatherman 50
R.R. 1 pt
Valley Center, Kansas 67147 po
Crops: wheat, alfalfa, 1st yr. peaches
Regular commercial channels

Wayne Clutter 720
Waldron, ft
Kansas 67150 ao
Crops: wheat and beef cattle
Quantities: 6,000 bushels
Livestock: 50 head beef
Regular commercial channels

Vincent John Hoheisel 240
2642 N. Edwards pt
Wichita, Kansas 67204 po
Crops: wheat, corn, beans, alfalfa
Quantities: 4000 bu. wheat, 2000 bu. corn and
beans
Regular commercial channels

KENTUCKY

Mrs. Lilly Tomlinson 70
Rt. 1 ft
Brooksville, Ky. 41004 po
Crops: eggs, chickens, vegetables, beef
Quantities: 175 doz. week, average 1000 lbs.
Livestock: chickens, beef and ducks
Roadside stand

Fertility Acres 69
The Organic Farm ft
Curd Edmunds ao
Rt. 1
Glasgow, Ky. 42144
Crops: strawberries, potatoes, peanuts, sweet
potatoes, and vegetables
Quantitites: 200-300 bu. each kind of potatoes,
1000-1500 gal. berries

Larry Perkins 75
Horse Cave, R.R. 3 ft
Ky. 42749 po
Crops: tobacco and forage
Livestock: cows
Regular commercial channels

John P. Cassady   30
P.O. Box 322   pt
Inez, Ky. 41224
Crops: hay
Quantities: 1 ton per acre
I market my harvest to other farmers.

John Keller   297
802 Chinoe Rd.   ft
Lexington, Ky. 40502
Crops: tobacco, corn
Livestock: beef cattle, sheep
Regular commercial channels

M & P Noll   4
Rt. 1, Box 13A   pt
Morning View, Ky. 41063   po
Crops: corn, beans, vegetables
Quantities: small

James Blackburn   40
RFD 5   pt
Princeton, Ky. 42445
Livestock: 8 head Hereford

Colonial Valley Veg Gardens   135
c/o Peggy Sue Turner   pt
R. 5
Winchester, Ky. 40391
Crops: tobacco and vegetables
Roadside stand

Mrs. John Hicks, Jr.   143
Rt. 4, Combs Ferry Rd.   pt
Winchester, Ky. 40391   po
Crops: hay, small garden, tobacco
Livestock: 32 Holstein feeder steers

LOUISIANA

Cleve Simon   12
Rt. 1, Box 145   pt
Basile, La. 70515   po
Crops: corn, okra, potatoes
Quantities: 40 bu. acre
Livestock: cattle, poultry

Del D. Hester   20
P.O. Box 400   pt
Jena, La. 71342   po
Crops: various vegetables
Livestock: 100 head hogs
Wholesaler
Regular commercial channels

Douglas Waddell   25
Rt. 1, Box 12-B   ft
Kinder, La. 70648   po
Crops: vegetables
Livestock: poultry, rabbits, hogs
Roadside stand and in our own private camp-
ground.

Casey Pierce
301 Birch Dr., Apt. 55
Lafayette, La. 70501   po
Crops: rice
Livestock: 80 head of cattle

J. M. Ingalls   95
Rt. 3, Box 85   ft
Leesville, La. 71446   po
Crops: soybeans, Lindsey 77, rye grass and
wheat for pastures
Livestock: 60 cows and yearlings
Regular commercial channels.

Elmo Brown   80
Rt. 2, Box 214   pt
Minden, La. 71055
Crops: vegetables

MAINE

Cliff McDonough   30
29 Main St.   pt
Belfast, Maine 04915   ao
Crops: garden produce
Livestock: 12 sheep, 2 beef cattle, hens, pony
Market my harvest to my customers in my
business.

Robert C. Morse, Jr.   ¼
Samoset Rd.   pt
Boothbay, Maine 04537   ao
Crops: peas, beans, beets, broccoli
I sell harvest to friends.

Mrs. Rose Lavallee   35
RFD 2
Bowdoinham, Maine 04008
Crops: blueberries, raspberries, blackberries,
rhubarb, elderberries and grapes

Mr. & Mrs. G. Rich   less than 10
Bradford,   ft
Maine 04410
Crops: vegetables for year round home use.

Miss Yoland Romano   14
RFD 2   pt
Chester, Maine 01011   ao
Crops: vegetables
Livestock: 2 cows

Howard L. Carr, Jr.   25
Box 103   pt
Cornish, Maine 04020   po
Crops: hay & family vegetables
Livestock: goats, pigs, calf, horses

Walter L. Harriman   1
Rt. 1   pt
Dixmont, Maine 04932   ao
Crops: potatoes, corn, beans
Livestock: 1 goat, 1 calf, 65 chickens

Kenneth Horn   1
RFD 1   pt
Dixmont, Maine 04932   ao
Crops: vegetables
Livestock: ducks
Roadside stand, direct to retail store.

Walter J. Ruginski   1
RD 3   pt
Gorham, Maine 04038   ao
Livestock: beef, goats, sheep and chickens

Warren & Denise Cochrane   30
Greenridge Organic Farm   pt
Greenville, Maine 04441   ao
Crops: various vegetables
Livestock: goats, pigs, chickens, sheep, turkeys,
ducks
Roadside stand
Regular commercial channels

Scott & Helen Nearing   4/5
Harborside,   ft
Maine 04642   ao
Crops: blueberries
Quantities: 1034 qts.
Direct to retail store.

Sandra Buker   40
Kennebunkport,   pt
Maine 04046   po
Crops: Timothy hay and garden produce
Livestock: 12 nubian goats

Phillip McBrien   50
Star Rt. 2   ft
Liberty, Maine 04949   ao
Crops: eggs, milk, vegetables
Quantities: 4 doz. eggs and 20 qts. a day
Livestock: cows, sheep, hens, rabbits, doves
Roadside stand

Sue Wilson   60
RFD, Box 135   pt
Lincoln Center, Maine 04458   po
Crops: beef
Quantities: 50 head yearly
Livestock: draft horses, pigs, dairy heifers
Regular commercial channels

Ellis Dodge   25
Dodge Rd.   pt
North Berwick, Maine 03906   ao
Crops: cattle and vegetables
Livestock: 15 head cattle

Eileen LeRoy   1
River Rd.   pt
N. Leeds, Main 04263   ao
Crops: vegetables for home use.
Livestock: hope to raise chickens and beef

Brian Dreher   12-18
R.R. 2
N. Whitefield, Maine 04353
Crops: vegetables, grains
Livestock: chickens, goats and sheep

David Brookes Pearson   75
175 Park St.   pt
Orono, Maine 04473   ao
Crops: vegetables, some grains next year.
I sell to friends and give away.

Jim Zukauskas   70
4 Schoppe's   ft
Orono, Maine 04473   ao
This will be our first season on the farm.

Robert Allen   40
Boothley Park   pt
Scarborough, Maine 04072   po
Crops: hay, strawberries, tomato, cukes
Livestock: 7 horses

Mrs. Sophie Knight   25
Rt. 1, Box 44   ft
Topsham, Maine 04086   ao
Crops: vegetables and fruits
Roadside stand

Edward Howe   50
RFD 1   pt
Union, Maine 04862   ao
Crops: vegetables, hay
Quantities: relatively small
Livestock: goats and chickens
Roadside stand

Belmont Smith   60
Unity,   pt
Maine 04988   po
Crops: strawberries, corn, garden vegetables
Livestock: 4 milk cows, 6 calves, 5 bulls, 6
heifers
Sell harvest to local people.

Goose River Farm   20
R.F.D. 3, Box 97   ft
Waldoboro, Maine 04572   ao
Sell to wholesaler, roadside stand and direct to
retail store.

William C. Pierce   30
West Baldwin,
Maine 04091
Crops: hay
Livestock: 4 grade Jersey cows

Richard Dren   2
1042 Bridgeton Rd.   pt
Westbrook, Maine 04092   ao
Crops: peas, potatoes, corn, lettuce, beans,
cukes, etc.
Quantities: 260 rows-300-400-50-150-25
Roadside stand

MARYLAND

Floyd W. Smith   102
Qtrs, C, NSRDC   pt
Annapolis, Md. 21402   po
Crops: beef, soybeans, corn
Livestock: beef only
Wholesaler and mail order
Regular commercial channels

Miss Fannie Harn   ½
4402 Roke by Road   ft
Baltimore, Md. 21229   ao
Crops: raspberries, herbs, sunflower seeds and
honey

Ronald A. Nelson   25
36 Canary Rd.   pt
College Park, Md. 20740   po
Crops: corn, tomatoes, beets
Quantities: 2 acres corn
Livestock: 6 head beef animals
Wholesaler
Regular commercial channels

Steve Hudnall   45
Rt. 2, Box 268   ft
Flintstone, Md. 21530   po
Crops: Jerusalem artichokes, sunflowers,
peanuts
Livestock: cattle and hogs

F. V. Smith   1
Flintstone,   pt
Md. 21530   ao
Crops: tomatoes
Quantities: about 7000 lbs.
Direct to retail store

Col. J. F. McClanahan   45
USMC (Ret)   pt
732 Earlton Rd.
Havre de Grace, Md. 21078
Crops: hay

Mrs. Virginia Rowe   20
2718 Kirkwood Pl.   ft
Hyattsville, Md. 20782   po
Crops: soybeans
Quantities: 200 bu.
Sell my harvest to grain market.
Regular commercial channels

Julian Morris   150
Marydel,   ft
Md. 21649   po
Crops: soybeans, corn, vegetables
Quantities: 100 acres soybeans, 5 acre of veg.
Livestock: chickens, geese, ducks, hogs
Regular commercial channels

Eli A. Yoder   200
R. 2, Box 129   ft
Oakland, Md. 21550   ao
Crops: corn, oats, hay and some vegetables
Quantities: 400 baskets corn, 3000 bu. oats, 100
tons hay
Livestock: 37 Holstein dairy cows, 20 heifers
Feed just about all my crops to the cattle.

E. S. Higgins   25
Rt. 1, Box 54   ft
Oldtown, Md. 21555   po
Crops: fruits and vegetables
Quantities: 500 to 1000 bu.
Livestock: 2 ponies
Regular commercial channels

Frances Ingles   2
Rhodesdale,   ft
Md. 21659   ao
Crops: turnips, kale, collards, crooked neck
pumpkins and beets
Livestock: few chickens
Direct to retail store, few mail orders and in-
dividuals.

Hugh H. Mercer   186
Oxford Ferry Rd.   Share Basis
Royal Oak, Md. 21662
Crops: corn, soybeans
Wholesaler
Regular commercial channels

White Wilder   11
Solomon,   ft
Md. 20688
Crops: garden and tobacco
Wholesaler and roadside stand

Carlton Barnes   85
Rt. 2, Box 312   ft
Sykesville, Md. 21784   ao
Crops: corn, wheat, barley, sweet corn, toma-
toes, potatoes
Quantities: 25 acres, 8, 8, 10, 1 and tomatoes
for DC market
Livestock: steers, cows, calves
Roadside stand and direct to retail store.

MASSACHUSETTS

Joseph Marques   20
644 Middle Rd.   ft
Acushnet, Mass. 02743   ao
Crops: garden vegetables, goats milk
Quantities: ½ acre corn, carrots
Livestock: 2 goats, 2 pigs/cows, 3 heifers, 25 chickens
Roadside stand and sell surplus.

Stan & Maryjane Bean   10
Orchard Park Farm   ft
Amesbury, Mass. 01913   ao
Crops: vegetables, blueberries, apples, strawberries, pears
Quantities: 250 qts. blueberries, 50-150 bu. apples
I sell my harvest to customers who order in advance.

Bruce Gustavsen   2
Butterhill Farm   pt
Amherst, Mass. 01002   ao
Livestock: 3 goats and 25 chickens

Peter Wartiainen, Jr.   25
Old Hardwick Rd.   pt
Barre, Mass. 01005   ao
Crops: poultry, pork, vegetables, potatoes, hay
Quantities: 150 tons potatoes, 20 bu. vegetables
Livestock: 50 hens, 100 meat chickens, 10 pigs
Retail at farm.

Penelope Turton   1½
Stearns Farm   ft
859 Edmands Rd.   ao
Framingham, Mass. 01701
Crops: all usual vegetables
Livestock: hens only
Roadside stand with my health food store.

Mark Kramer   11
Catamount Hill Farm   ao
Griswoldville, Mass. 01345
Crops: hay and tomatoes
Livestock: pigs and cows
Direct to retail store
I market my harvest through personal contacts.

Stanley Armstrong   6
Arrowhead Springs   pt
   Organic Farm   ao
174 Center St.
Groveland, Mass. 01830
Crops: veg. plants, honey, table vegetables
Roadside stand

Robert O. Domin   15
Millvale Farm & Cider Mill   ft
85 Millvale Rd.   ao
Haverhill, Mass. 01830
Crops: corn
Roadside stand

Daniel Griffith   10
Mumford Hill   pt
Manchaug, Mass. 01526   ao
Plan to plant fruit trees, berries and establishing a garden.

Stanley Lowell   30
Popes Pt. Rd., RFD 2   ao
Middleboro, Mass. 02346
Crops: asparagus
Quantities: 2 acres

R. M. Alberti   1
Rt. 1, Box 165   pt
Vining Hill Rd.   ao
Southwick, Mass. 01077
Crops: tomatoes, beans, carrots, peas, pickles, onions, potatoes, apples
Quantities: 50 qt., 30 qt., 30 pt., 30 qt., 30 qt., 3 bu., 8 bu., 8 bu.
Livestock: 100 chickens, 3 sheep
Roadside stand
Regular commercial channels

Gaston Plaguet   3
Sergeant St., P.O. Box 255   pt
Stockbridge, Mass. 01262   ao
Crops: vegetables, fruits
Quantities: 25 bu. vegetables
Plan to open stand and also buy from other producers.

Mrs. Joseph Keith III   12
775 Horseneck Rd.   pt
Westport, Mass. 02790   po
Crops: peas, beans, beets, carrots, parsley, squash, etc.
Quantities: 10 bushel peas
Livestock: 2 ponies, expect 2 sheep and a beef cow

Mrs. Arnold Bossi   ¼
Keveney Lane   pt
Yarmouth Port, Mass. 02675   ao
Crops: vegetables
Quantities: not much
Livestock: 2 dairy goats

MICHIGAN

Dorothy Dickerson   67
28611 D Drive N.   pt
Albion, Mich. 49224   ao
Crops: grains and vegetables
Livestock: beef cattle, goats, chickens
I sell my harvest to the organic market through interested friends.

Dennis Dellaporte   30
6140 Swartout Rd.   pt
Algonac, Mich. 48001   ao
Livestock: ducks, geese, turkeys, chickens, guineas

Fred Zaika   140
Rt. 2   ft
Bearlake, Mich. 49614   po
Crops: asparagus, strawberries, hay
Quantities: 60 tons or more
Regular commercial channels

McCraney's Big Hill Farm   40
R. 2, Box 204   pt
Boyne City, Mich. 49712   ao
Crops: table queen squash
Quantities: many bushels
Livestock: steers and rabbits
Roadside stand

Jack Wheaton   30
Box 136   pt
Brighton, Mich. 48116   po
Crops: hay & small vegetable garden
Quantities: 24 tons of hay
Livestock: 2 donkeys

Mrs. Alfred Schuh   10
Rt. 2, Box 416   pt
Buchanan, Mich. 49107   ao
Crops: sweet corn, peaches
Quantities: 4 acres sweet corn, 1 acre peaches
Livestock: chickens-100 a year
Roadside stand

John Dickerson   155
Rt. 2   ft
Cassopolis, Mich. 49031   ao
Crops: calves
Livestock: 40 cows, 4 horses
Market my harvest to local feeders.

Mary Ann Dedenbach   ¼
Rt. 2   ao
Cheboygan, Mich. 49721
Crops: carrots, parsnips, peas, corn, beets
Quantities: 6 bu. beans, 8 bu. corn
Livestock: 50 chickens

Maynard Allison   30
Clarksville,   pt
Mich. 48815   po
Crops: beans, corn
Quantities: 6 to 10 acres
Livestock: sheep
Regular commercial channels

Joseph A. Yoder   180
R. 2   pt
Constantine, Mich. 49042   po
Crops: oats, alfalfa, corn, soybeans, beef
Quantities: 40, 40, 50, 50 acres
Livestock: beef, hogs and goats
Regular commercial channels

Donna Hanson 40
15323 Burgess St. pt
Detroit, Mich. 48223 ao
Crops: fruit trees

D. H. Cooper 50
6330 Daly Rd. pt
Dexter, Mich. 48130 po
Livestock: horses, sheep, chickens, cattle
Regular commercial channels

J. Rostafinski 1
724 Withington pt
Ferndale, Mich. 48220 ao
Crops: corn, tomatoes, beans, onions, beets,
honey
Sell my harvest to neighbors and friends.

Roger Staples 240
Germfask, ft
Mich. 49836 ao
Livestock: deer, rabbits, birds, grass

George Walsh 40
2045 Emerson Rd. pt
Goodells, Mich. 48028 ao
Livestock: pigs, chickens, cow and horse

Walter Wheeler 100
Rt. 3 pt
Hart, Mich. 49420
Crops: wheat, rye, sweet corn, tomatoes,
strawberries
Livestock: cattle, pigs
Regular commercial channels

Louis H. Spears 50
Rt. 1, Box 77 pt
Hersey, Mich. 49639 po
Crops: hay
Quantities: 600 bales
Livestock: 10 feeder calves

Ron & Genevieve Alexander 75
6030 Hunters Creek Rd. pt
Imlay City, Mich. 48444 po
Crops: wheat, oats, hay
Quantities: 120 bu. wheat, 400 bu. oats and
2000 bales hay
Livestock: 30 head, 200 chickens

Moses Bender 200
R. 2 pt
Ithaca, Mich. 48847 po
Crops: corn, beans, small grains
Livestock: poultry, hogs
Regular commercial channels

Edward Smith
Rt. 1, Box 231 pt
Lawrence, Mich. 49064 po
Crops: asparagus
Quantities: 1 acre

Stan Raichel 26
5304 W. Ray Rd. pt
Linden, Mich. 48451 ao
Crops: hay & corn
Livestock: beef, chicken, wild geese, pheasant
and fish

Jack Greenwald 6
R.R. 3, Box 270-A pt
Lowell, Mich. 49331 ao
Crops: vegetables and fruit
Direct to retail store
Regular commercial channels

Harley Galer 40
McCosta, pt
Mich. 49332 ao
Crops: gardening berries
Roadside stand

Harold Klais 40
1428 Georgia pt
Marysville, Mich. 48040 po
Crops: corn, pumpkins, sugar beets
Quantities: 25 acres
Livestock: cows, chickens
Wholesaler, roadside stand
Regular commercial channels

T & S De Haan 7
1143 144th St. ft
Moline, Mich. 49335 ao
Crops: beans, tomatoes, peas
Quantities: 700 bu. beans, 25 tons tomatoes and
200 bu. peas
Livestock: 10 goats and 10 chickens
Regular commercial channels

Joe H. Choate 30
R.R. 2, Box 50 A
Nashville, Mich. 49073

Buell Crosby 10
6241 Lake Pleasant Rd. pt
North Branch, Mich. 48461 ao
Crops: garden
Livestock: 5 head cattle
Farm Market

Mrs. Glen Jackson 30
1142 Bird Lake Rd. pt
Ossco, Mich. 49266 ao
Crops: rye, vegetables, milk, meats, eggs
Livestock: 5 cows, 3 sheep, 4 heifers, 25 banties,
rabbits

Helen Van Dyke 20
9585 Joy Rd. pt
Plymouth, Mich. 48170
I am interested in putting my fields into organic
grown produce, but don't know how to start.
Fields are in grass now.
Livestock: riding horse, few chickens

Dwight L. Havens 25
164 S. Fremont pt
Rockford, Mich. 49347 ao
Crops: vegetables

James Doty 27
Rt. 2 pt
Sheridan, Mich. 48884 ao
Crops: berries and garden produce
Livestock: 2 draft horses
Sell door to door.

Dale Crandall 35
Webberville, ft
Mich. 48892 po
Crops: oats, wheat, corn
Livestock: 500 hens
Roadside stand

Ed Makielski Berry Farm 61
7130 Platt Rd. ft
Ypsilanti, Mich. 48197 po
Crops: strawberries, raspberries, blackberries,
currants, gooseberries
Quantities: 36,000 qts. strawberries, 12,000
qts. raspberries

MINNESOTA

Norman Syzsiner 200
Alexandria, ft
Minn. 56308 ao
Crops: milk, beef, wheat, garden crops
Livestock: 30 milk cows, 20 beef animals
We have a natural food store on our farm.
Milk I market on regular commercial channels.

Wayne Swisher 130
Rt. pt
Badger, Minn. 56714
Crops: hay & pasture for milk & beef
Regular commercial channels

Harlan Riggs 109
Bagley, ft
Minn. 56621
Crops: potatoes, oats, hay and some wheat
Quantities: 20 acres potatoes, 15 bales hay and
oats
Livestock: 1 cow, 100 hens

Mr. & Mrs. A. Sheria 45
P.O. Box 27
Bigford, Minn. 56628
Crops: vegetables & hay

Leonard Genzler   200
379 S. Pine St.   ft
Caledonia, Minn. 55921   ao
Crops: wheat, asparagus, peas, sweet corn, berries
Quantities: 10 acres, 1 acre, 30 acres, 30 acres, 1 acre
Livestock: 35 holstein cows plus calves
Sell to consumers

Jerry McEvilly   80
R.R. 2   pt
Caledonia, Minn. 55921   po
Crops: cucumbers, squash, will have fruits and berries
Livestock: organic lambs
I sell my harvest to individuals.
Regular commercial channels

Roger Ekstrand   220
Dawson,   pt
Minn. 56232
Crops: oats, corn, alfalfa, pasture
Livestock: sheep
Regular commercial channels

Erwin Moen   195
Deer Creek, Minn. 56527   pt
Crops: corn, grain   po
Quantities: 2,000 bu.

Peter Timmers   10
R.R. 1   pt
Dendas, Minn. 55019   po
Crops: corn, grain, alfalfa
Livestock, beef, veal, chickens
Wholesaler
Regular commercial channels

Duane R. Shellum   12
Chozen Acres   pt
Box 97A Route   ao
Elko, Minn. 55020
Crops: not decided
Livestock: chickens, ducks, geese, donkeys, ponies

Lyle D. Schleshe   100
RR 2   ft
Erhard, Minn. 56534
Crops: oats, wheat
Quantities: 25 bu. to the acre
Livestock: swine, beef
Regular commercial channels

Oliver Kramer   300
R.R. 2   ft
Gibbon, Minn. 55335   ao
Crops: corn, soybeans, alfalfa, oats
Quantities: 40 acres corn, 100 soybeans, 60 alfalfa and 60 oats
Livestock: dairy, dairy beef, chickens
Regular commercial channels

Wayne H. Daak   90
R.R. 2   pt
Glencoe, Minn. 55336
Crops: corn, oats, soybeans
Quantities: 26 acres corn, 31 acres beans, 16 acres oats
Regular commercial channels

Robert Sundre   55
R.R. 2   pt
Glenville, Minn. 56036
Crops: hay, soybean, corn, raspberries
Quantities: ¾ acre of raspberries
Livestock: hogs, sheep
Regular commercial channels

Lawrence A. Swanson   35
Rt. 3, Box 255   ft
Grand Rapids, Minn. 55744   ao
Crops: strawberries, raspberries, rhubarb, asparagus

Harold W. Zimmerman   3
926 W. 3rd   pt
Hastings, Minn. 55035   ao
Crops: potatoes, sweet corn, squash and melons
Quantities: ½ acre of each
Roadside stand and direct to retail store.

Rodney's Organic   130
  Beef Supply   pt
Rt. 1, Box 8   ao
Henning, Minn. 56551
Crops: beef, wheat
Livestock: Angus beef and some Hereford beef
Direct to retail store and consumer.

W. M. Wilharber   83
6949 Centerville Rd.   pt
Hugo, Minn. 55038   po
Crops: assorted vegetables
Quantities: 20 bu. tomatoes
Livestock: 6 horses

Herbert White   40
Rt. 2   pt
Kimball, Minn. 55353   ao
Crops: sweet corn
Livestock: 3 beef, 25 hens and 5 turkeys
Roadside stand

Gerhardt Kohrs   160
R.R. 3, Box 142   ft
Lake City, Minn. 55041   ao
Crops: wheat, oats, corn, alfalfa
Livestock: dairy, beef, hogs
Regular commercial channels

Mrs. Ken Hills   20
Rt. 1, Box 33
Meadowlands, Minn. 55765
Crops: vegetable garden
Livestock: will have 100 chickens

Jerome Kelly
8731 Sheriden Ave. S   pt
Minneapolis, Minn. 55431   ao
Crops: rabbits

Lester Frohrip   320
Morgan,   ft
Minn. 56266   po
Crops: corn, beans, potatoes, alfalfa, oats, carrots
Livestock: cattle and hogs
Regular commercial channels

Eino Hill   100
Box 34   pt
Mt. Iron, Minn. 55768   ao
Crops: hay, grain
Quantities: 500 bu. grain, 40 tons hay
Livestock: chickens
I sell direct to consumer.

Celesten Champoux   100
Northome,   ft
Minn. 56661   ao
Livestock: pigs
Not in production yet.

Mr. & Mrs. Milton Teufer   60
Northome,   pt
Minn. 56661   ao
Crops: hay
Quantities: 800 bales hay
Livestock: cattle, ducks, chickens, wild ducks, geese, turkeys
Regular commercial channels

Neil Colbenson   140
Box 56   ft
Peterson, Minn. 55962   ao
Crops: oats and hay
Livestock: beef, dairy, chickens
Regular commercial channels

Glen Sherwood   25
Longville Rt.   pt
Pine River, Minn. 56474   ao
Have just moved to the farm and will begin operation in the spring.
Roadside stand

Clifford Lundholm   120
Rt. 1   ft
St. Peter, Minn. 56082   ao
Crops: corn, oats, alfalfa
Livestock: dairy herd about 50
Regular commercial channels

Mrs. Herman C. Kjos   151
403 State Ave. N.   ft
Thief River Falls, Minn. 56701
Crops: wheat, barley, oats
Regular commercial channels

MISSISSIPPI

Robert Finley   10-40
Rust College   ft
Holly Spring, Miss. 38635   ao
Crops: sunflower seeds
Quantities: 10-20 acres
Wholesaler

William Morris   40
Rt. 2, Box 158-A   pt
Summit, Miss. 39666   po
Crops: tomatoes, beans, cukes, squash, corn
Roadside stand and direct to retail store.

G. F. McGreger   35
531 N. Robins
Tupelo, Miss. 38801
Expect to get started working on it in 1971 and
be ready for full effort in 1972.

MISSOURI

Leo J. Lindsey   200
Rt. 1   ft
Burnett, Mo. 65011   ao
Crops: corn, wheat, soybeans, oats
Quantities: 30 acres of each
Livestock: cattle, goats, chickens, horses
Wholesaler, roadside stand, direct to retail
store and by mail.
Regular commercial channels

Ethel Sporich   188
Box 162   pt
DeSoto, Mo. 63020   ao
Crops: vegetables, alfalfa
Livestock: poultry & livestock

Wilbert Krueger   112
R. 3   pt
Ellington, Mo. 63638   po
Crops: hay and pasture
Quantities: 2000 bales hay
Livestock: 25 calves
Regular commercial channels

Joseph Harfmann   30
308 S. 2nd St.   ft
Elsberry, Mo. 63343   ao
Crops: beef
Livestock: hogs and chickens
Roadside stand

Arthur N. Buss   377
Box 117
Eolia, Mo. 63344
Converting all my acres to grass and legumes
for beef production.

Al Mueller   14
233 S. Dade   ft
Ferguson, Mo. 63135   ao
Crops: tomatoes, sweet potatoes, cabbage, con-
cord grapes
Quantities: by the truck load
Wholesaler, roadside stand
Regular commercial channels

Lyle R. Allen   40
Star Route, Box 22   ft
Fredericktown, Mo. 63645   ao
Crops: tomatoes, other vegetables
Livestock: 11 sows
Wholesaler and roadside stand
Regular commercial channels

O. D. Draper   40
2514 Race   pt
Independence, Mo. 64052   ao
Crops: tomatoes, corn
Direct to retail store, roadside stand

Paul R. Bennett   30
2305 Ozark
Joplin, Mo. 64801
Crops: beef
Livestock: 12 head of cattle

Jim Riffle   40
Kahoka,   pt
Mo. 63445   ao
Crops: sweet corn, tomatoes, potatoes
Quantities: 20 acres corn, 10 acres tomatoes,
7 acres potatoes
Livestock: 5 steers, 10 pigs, 2 horses
I own store.

L. V. Cannon   40
5907 E. 100th St.   pt
Kansas City, Mo. 64134
Crops: grass & garden products, fruit

William Litz   90
14200 Peterson Rd.   pt
Kansas City, Mo. 64149   ao
Crops: alfalfa and garden vegetables
Livestock: 31 horses, 2 goats, 100 rabbits, 2
turkeys, 4 chickens and more coming

Mrs. Kenneth Peterson   35
Rt. 1   pt
Kingsville, Mo. 64061   po
Crops: organic eggs
Quantities: 100 doz. eggs week
Livestock: 500 hens
Direct to retail store

Victor Kirschner   50
R. 1, Box 33   pt
Lonesdell, Mo. 63060   ao
Crops: meat, garden and eggs
Livestock: cattle, hogs and chickens

Ralph Filbeck   40
Marionville,   pt
Mo. 65705   ao
Crops: wheat, beef cattle
Livestock: 13 cows, 1 bull, 10 calves
Regular commercial channels

Gertrude I. Lucas   98
R. 2   ft
Marshall, Mo. 65340   po
Crops: corn, soybeans and wheat
Quantities: 8000 bu. corn, 1200 beans, 1000
wheat
Regular commercial channels

O. K. Hamlett   110
Rt. 1, Box 126   ft
Neelyville, Mo. 63954   po
Crops: soybeans, corn, fruit & vegetables
Livestock: hogs and poultry
Regular commercial channels

Elmer C. Eickelmann   66
R.R. 2   pt
Owensville, Mo. 65066   ao
Crops: beef, feed grain and hay
Livestock: 2 hereford cows and 100 chickens

Gerald A. Maas   40
Rt. 1   ft
Poplar Bluff, Mo. 63901   ao
Crops: corn and hay
Livestock: hogs and chickens
Regular commercial channels

Dr. & Mrs. Leland J. Dean   33
207 N. Washington   pt
Princeton, Mo. 64673   ao
Crops: hay
Livestock: beef
Wholesaler
Regular commercial channels

Viola M. Bradley   12
Rt. 1   pt
Revely, Mo. 63070   po
Crops: tomatoes and bush beans
I give my harvest away and for family use.

96

Hugh Glackin 200
Rt. 1 ft
Richland, Mo. 65556 ao
Crops: hay and pasture
Quantities: 20 beef cattle
Livestock: cattle and cows
Regular commercial channels

Lawrence Butler 30
R.R. 2, Box 71 pt
Siligman, Mo. 65745 ao
Crops: garden and pasture
Livestock: 10 cows, 6 hogs
Regular commercial channels

Ray F. Bales 118
Stet, pt
Mo. 64680
Crops: corn, oats, wheat, clover and grass land.
Quantities: 20 acres corn, 20 acres oats, 8 acres
wheat, 20 acres clover
Livestock: cattle, hogs, poultry
Regular commercial channels

Flavious D. Adkins 22
Rt. 1, Box 221 pt
Troy, Mo. 63379 ao
Crops: corn
Livestock: 5 horses and 3 cows
Regular commercial channels

Larry Dodd 30
RR 2, Box 52 pt
Troy, Mo. 63379 po
Crops: pasture
Livestock: cattle, pigs, rabbits

Ralph Barnhart 20
Rt. 1 ft
Warrensburg, Mo. 64093 ao
Crops: soybeans, alfalfa
Quantities: 450 bu.
Livestock: beef on grass and hay
Wholesaler
Regular commercial channels

MONTANA

Mel Gore 20
417 North 3rd
Bozeman, Mont. 59715
Interested in organic gardening and farming.

John J. Earley 75
Star Rt. 6 ft
Broadus, Mont. 59317 ao
Quantities: 5,000 lbs. alfalfa seed
Crops: lambs & alfalfa seed
Livestock: 36,000 lbs. lamb
Regular commercial channels

V. R. Johnson 1200
Box 63 ft
Homestead, Mont. 59242 ao
Crops: wheat, oats and rye

Terra Floranna Industries 70
Douglas Barrie, Gen. Mgr. ft
P.O. Box 721 ao
Kalispell, Mont. 59901
Crops: grains, hay, beef
Quantities: 80-100 bu. p.ac. wheat, 100-120 bu.
p.ac. barley, 4-6 ton p.ac. hay
Direct to retail store
Regular commercial channels

Dale Heath 1440
311 8th Ave. S. ft
Shelby, Mont. 59474 po
Crops: wheat, barley, millet
Quantities: 20,000 bu. wheat, 15,000 bu. barley
5,000 bu. millet
Regular commercial channels

Mrs. Phyllis Falconer 40
Box 8 ft
Trego, Mont. 59934 ao
Crops: hay, pasture
Livestock: 6 horses

NEBRASKA

Mrs. Matt Zmek 40
Archer, ft
Nebr. 68816 po
Crops: corn, oats
Quantities: 5000 bu.
Livestock: Black Angus cattle, laying hens
Regular commercial channels

Dennis DeBlau 104
Fordyce, ft
Nebr. 68736
Crops: corn, alfalfa, rye
Quantities: 800 bu. corn
Livestock: cattle & sheep, chickens
Regular commercial channels

Gaylord Cradduck 80
Ohiowa, ft
Nebr. 68416 po
Crops: corn and milo
Livestock: hogs
Regular commercial channels

I. W. Hunsberger 35
Rt. 3 ft
Ravenna, Nebr. 68869 po
Crops: 45 acres pasture, 35 acres farming
Livestock: 25 head sheep

B. F. Kroeker 80
Steele City, ft
Nebr. 68440 po
Crops: potatoes and onions and most all veg.
Quantities: 100 bu. potatoes
Livestock: 14 milk cows and calves
Roadside stand

Howard Hohlen 25
Poeny Gardens pt
Trumbull, Nebr. 68980 po
Crops: corn, onions
Quantities: 600 bu. corn
Livestock: 30 head of sheep
Regular commercial channels

John H. Rasmus 120
Rt. 3 ft
West Point, Nebr. 68788 ao
Crops: corn, wheat, alfalfa and potatoes
Quantities: 450 bu. wheat, 2500 bu. corn
Livestock: few hogs and chickens
Regular commercial channels

NEVADA

Glenn Carle 80
Whippletree Farm pt
Farm Dist. Rd. ao
Fernley, Nevada 89408
Crops: beef, and vegetables
Livestock: beef, sheep, chickens 25 head
Roadside stand and direct to retail store
Regular commercial channels

NEW HAMPSHIRE

Mrs. Harold Bemis 1
RFD pt
Charlestown, N. Hamp. 03603 ao
Crops: for home use
Livestock: steers, pigs, chickens

John A. Buote 110
Farmington, pt
N. Hamp. 03835 po
Crops: hay, vegetables
Livestock: 1 goat
Wholesaler
Regular commercial channels

Timothy Jones ½
RFD 1 pt
Madbury, N. Hamp. 03820 ao
No main crops
Regular commercial channels

George Wason 32
Box 148 pt
New Boston, N. Hamp. 03070 po
Crops: squash, tomatoes, potatoes and glads
Quantities: 1500 glads, 100 bu. tomatoes, 30
bu. potatoes, 2 tons squash
Livestock: 50 hens
Regular commercial channels

Mary L. Bates 1
Dana Hill Rd. pt
New Hampton, po
N. Hamp. 03256
Crops: vegetables
Direct to retail store

Col. Ethan R. Pearson 50
Garrison Organic Farm ft
Bay Rd. ao
Newmarket, N. Hamp. 03857
Crops: home garden and fruits
Livestock: beef, sheep, poultry
Regular commercial channels

Lowell E. Rheinheimer 15
Thyme Hill Farm ft
Temple, N. Hamp. 03084 ao
Crops: apples
Quantities: 5,000 bushels

NEW JERSEY

Garden of Ilsley 8
H. Wallace Ilsley ft
RD 1, Box 265A ao
Asbury, N.J. 08802
Crops: fruit, berries, vegetables
Livestock: sheep, poultry
Roadside stand

Nicholas R. Anzevino 50
RD 1, Box 334 pt
Boonton, N.J. 07005 ao
Crops: hay and shrubs
Quantities: 2,000 bales, 500 trees
Sell my harvest to feed dealer.

John Himich 27
701 Cranbury Rd. pt
East Brunswick, N.J. 08816 po
Crops: apples, vegetables
Quantities: 500 bu. apples, 4 acres veg.
Direct to retail store and interested parties.

Richard Zaleski 120
PeaPacton Farm ft
Box 68 po
Far Hills, N.J. 07931
Crops: beef, hay
Livestock: horses and steers
Regular commercial channels

William Power 28
RD 2, Box 192 pt
Flemington, N.J. 08822 ao
Crops: hay, corn

Mr. & Mrs. Charles Tickner 7
RD 2 pt
Hackettstown, N.J. 07840 po
Crops: tomatoes, squash, cucumbers, carrots
Livestock: 2 bulls
I use harvest myself at present, will have a
store this year or next.

Albert N. Olsen 97
Walnut Hill Farm pt
RD 2, Mountainville po
Lebanon, N.J. 08833
Crops: corn, vegetables, hay
Quantities: 14 acres corn, 12 acres hay, ½
acre of veg.
Livestock: 3 ponies, 3 beef cattle, 10 hens, 17
meat-type roosters

Edward Leidell 7
1553 W. Front St. pt
Lincroft, N.J. 07738 ao
Crops: vegetables, fruit, eggs, milk
Livestock: 2 cows, goats
Sell my harvest to customers who come to
house.

Edward J. Forster 35
698 Worth Ave. pt
Linden, N.J. 07036 po
Would be very interested in complete organic
farming—eggs, fruits and vegetables.
Livestock: 10,000 chickens
Wholesaler and direct to retail store
Regular commercial channels

Dale Hurliman 112
Full Circle Farm ft
Long Hill Rd. ao
Neshanic Station, N.J. 08853
Crops: apples, peaches, & pears
Quantities: 20,000 bu. apples & smaller
amounts of other fruit
Wholesaler and later retail some ourselves.
We will be in business for 1971 crop year.

Joseph Golden 30
449 W. Grand Ave.
Rahway, N.J. 07065
To begin farming in the near future.

Karl Myers 29
508 Thornau Dr.
Rivervale, N.J. 07675

Robert Redrath, Jr. 75
423 N. Ridgewood Rd. ft
S. Orange, N.J. 07079 po
Crops: hay and corn
Livestock: cows

Albert Idakaar 27
14 Chesler Square
Succasunna, N.J. 07876 po
Crops: corn, blueberries

Dennis McCaffrey 5
1452 W. Forest Grove Rd. pt
Vineland, N.J. 08360 ao
Crops: corn, tomatoes, beans, onions, cucum-
bers
Livestock: goats, ducks
Roadside stand

NEW MEXICO

Darrel Randall 40
1614 Gold S.E. ao
Albuquerque, N. Mex. 87106
Crops: alfalfa, honey. vegetables
Livestock: goats, chickens, ducks, geese, bees

John A. Nance 46
North Star Rt., Box 284 pt
Corrales, N. Mex. 87048 po
Crops: apples, pears, cherries
Quantities: 2,000 to 4,000 bu.
Wholesaler, roadside stand and direct to retail
store
Regular commercial channels

Dilia Cooperative Exchange 56
Box 19, Dilia Rt. ft
La Loma, N. Mex. 87724 ao
Crops: vegetables, corn, wheat, apples, pears
Livestock: goats, rabbits, chickens
Direct to retail store

NEW YORK

Richard Mathews 30
Amenia, pt
N.Y. 12501 po
Crops: hay, veg. pasture
Quantities: 10,000 bu. hay
Livestock: chickens, beef cattle, sheep
Regular commercial channels

Paul Kruger 40
1513 Burns Rd. pt
Angola, N.Y. 14006 po
Crops: hay, tomatoes
Livestock: beef, chickens
I sell my harvest through personal sales at farm.

98

Edward Karakajan   4½
RD 1, Conners Rd.   pt
Baldwinsville, N.Y. 13027   ao
Crops: eggs, vegetables, herbs, strawberries
Livestock: ducks and chickens
I sell my harvest at home.

Mrs. Arthur Woolson   5
Brocton,   ft
N.Y. 14716   ao
Crops: tomatoes
Quantities: 2000 gals. tomato juice
Livestock: 1 goat
Roadside stand, direct to retail store and at a
farmers market.

Roger Nevinger   25
Brookside, Rt. 2   pt
Adams, N.Y. 13605   ao
Crops: garden vegetables, fruit, hay
Quantities: enough for home use and friends

Richard Ryerson   11
RD 2   ao
Cambridge, N.Y. 12816
Livestock: goats and chickens

Sheryl M. Farnham   90
Ellicottville,   ft
N.Y. 14731   ao
Crops: corn, oats, potatoes
Quantities: 300 bu. corn, 500 bu. oats and 20
bu. potato
Livestock: beef cattle, poultry, horses
Wholesaler and local dealers

Diane Waltz   80
RD 1, Creek Rd.   pt
Esperance, N.Y. 12066   po
Crops: hay and vegetables
Livestock: sheep, rabbits, Herefords and
chickens

P. J. Mason
29 Heatherwood Rd.   ft
Fairport, N.Y. 14450   ao
Interested in starting a small organic farm.

Z. K. Elvi   10
Fishes Eddy,   ft
N.Y. 13774   ao
Crops: corn, tomatoes, carrots, beets, string
beans

Costello Organic   27
  Farm Equip.   pt
219 Berry Rd.   ao
Fredonia, N.Y. 14063
Crops: fruit and vegetables
Roadside stand

Deer Valley Farm   95
RD 1   ao
Guilford, N.Y. 13780
Crops: grain, beef, eggs, poultry, pork
Livestock: beef, poultry, hogs, dairy
We have our own outlet store.

John Bray   5
Mill St.   pt
Hannibal, N.Y. 13074   ao
Crops: potatoes, tomatoes, raspberries, pears,
string beans, peas, etc.
Quantities: 20 bu. potatoes, 100 qts. raspber-
ries, 20 or 30 bu. fruit
Livestock: pigs, chickens, horses & goats
Roadside stand, direct to retail store and
personal contact.

Miss Helen Ashworth   25
Rt. 2, Box 33
Heuvelton, N.Y. 13654
Crops: tomatoes, squash, strawberries, potatoes
& hay
Livestock: small herd of dairy cattle
Roadside stand

Fenton W. Nash   70
RD 1   ft
Himrod, N.Y. 14842   ao
Crops: grapes, grain, hay
Quantities: 15 tons grapes, 30 tons grain, 100
tons hay
Livestock: 24 beef
Wholesaler

Donald Haponski   2
RD 2   pt
Ilion, N.Y. 13357   ao
Crops: vegetables
Livestock: 3 horses

Mrs. T. H. Canfield   103
128 Eddy St.   pt
Ithaca, N.Y. 14850
Just purchased land and hope to farm organically.

Oliver Wendel Douglas   15
1019 Reynolds Rd.   pt
Johnson City, N.Y. 13790   po
Crops: tomatoes, garlic, onions
Quantities 6 bushels each
Livestock: chickens, pigs
Roadside stand

Nicholas Veeder   35
RD 1   pt
Jordanville, N.Y. 13361   ao
Crops: replacement heifers
Quantities: 10-12 a year
Livestock: chickens
Market heifers to other dairymen.

Elizabeth Darrah   35
Box 505   pt
Little York, N.Y. 13087   ao
Crops: apples
Quantities: about 200 bu.

Mack Farms   180
12551 Roosevelt Hwy.   pt
Lyndonville, N.Y. 14098   po
Crops: hay
Quantities: 35 ton
Livestock: registered hereford cattle

Robert Hartnagel   75
R. 1   ft
Lyons, N.Y. 14489   ao
Crops: wheat, oats and hay
Livestock: milk cows, beef cows, pigs, mink
Regular commercial channels

Kenneth Rice   100
Lyons,   ft
N.Y. 14489   po
Crops: cherries, apples, potatoes, beans
Quantities: 10,000 apples, 10 tons cherries
Regular commercial channels

George Stroebel   35
6999 McKay Rd., RD 2   pt
Mayville, N.Y. 14757   ao

W. K. Newman   ½
3 Dunn Rd.   pt
Monsey, N.Y. 10952   ao
Crops: strawberries

C. M. Einhorn   1⅓
RD 2   pt
Montgomery, N.Y. 12549   ao
Crops: vegetables and herbs
Livestock: 5 hens

James Frisch   50
Oak Hill Rd., RD 2   pt
Moravia, N.Y. 13118   ao
Crops: buckwheat
Quantities: 10,000 to 45,000 lbs.
Regular commercial channels

E. Kevane   60
Empire State Bldg.   pt
Suite 6515
350 Fifth Ave.
New York, N.Y. 10001
Livestock: pasture cattle
Interested in organic farming.

Kempton Smith   40
Manhattan Institute   pt
111 E. 31st St.   ao
New York, N.Y. 10016
Crops: vegetables
Livestock: beef cattle, goats, chickens

99

William G. Christopher   25
RD 1, Box 196   ft
Newark Valley, N.Y. 13811   ao
Crops: Just bought and will farm in spring
Livestock: goats, rabbits, chickens

William A. Leonard   30
Cadis Stage   pt
Oswego, N.Y. 13827   po
Crops: strawberries and vegetables
Livestock: 5 cows, 3 calves
Roadside stand

Mrs. Peter Ransom   20
Box 153   pt
Prospect, N.Y. 13435   ao
Crops: hay, corn and vegetables
Livestock: chickens

E. N. Hesford   50
RD 2   pt
Pulaski, N.Y. 13142   po
Crops: hay, corn, vegetables
Quantities: 3000 bales hay, 5 acres corn
Livestock: 11 head beef, 5 head hogs, 200
pullets
Roadside stand

P. H. Markle   26
RD, Enterprise Rd.   pt
Rhinebeck, N.Y. 12572   ao
Crops: hay, some vegetables
Livestock: sheep

Mrs. Kenneth Jones   48
102 Cobb Terr.
Rochester, N.Y. 14620
Interest in growing all crops organically.

Richard L. Carston   2
RD 2   pt
Salem, N.Y. 12865   ao
Crops: tomatoes, corn, squash, turnip, cabbage,
peppers
Quantities: ½ ton of each

Donald Pitts   25
Allen Rd., RFD 2   ft
Savannah, N.Y. 13146   ao
Crops: all kinds vegetables
Roadside stand

Donald L. Cross   25
Hanford Rd.   pt
Silver Creek, N.Y. 14136   po
Crops: vegetables
Sell to our friends and for own use.

Frank N. Carucci   45
67 Newark Ave.   ft
Staten Island, N.Y. 10302   ao
Crops: beets, cabbage, celery, corn, carrots
onions, cauliflower, sunflower seeds
Regular commercial channels

Harold H. Cooper   14
Box 313   pt
Warrensburg, N.Y. 12885   ao
Crops: eggs
Quantities: 10 doz. daily
Livestock: 100 cross bred hens & 100 araucanas

David E. Hull   35
Applewood Orchards   pt
Warwick, N.Y. 10990   po
Crops: apples
Quantities: 15,000 bu.
Wholesaler, direct to retail store and at the
farm.
Regular commercial channels and health food
stores.

Walter K. Cole   85
RD 4   pt
Watertown, N.Y. 13601   po
Crops: hay and grain
Livestock: cattle and poultry
Regular commercial channels

Thomas O. Owen   110
3115 Wesley Rd.   ft
W. Bloomfield, N.Y. 14585
Crops: grain, corn, alfalfa
Livestock: Holstein dairy—55 head
Regular commercial channels

Mrs. C. Perryman   42
22 Harvest Lane   pt
W. Islip, N.Y. 11795   po
Crops: vegetables
Livestock: 1 horse, 3 pigs, 50 chickens
Roadside stand

Mr. & Mrs. Leo Payne   50
Star Rt.   pt
Whitesville, N.Y. 14897   ao
Crops: garden products
Quantities: 20 bu.
Livestock: mostly chickens, poultry

Lewis Sellinger   72
County Line Farm   pt
Box 216A   ao
Woodbourne, N.Y. 12788
Crops: veg., tomatoes, cucumbers, squash
Livestock: chickens
Direct to retail store

NORTH CAROLINA

Harold Cory   60
2 Pine Tree Rd.   pt
Asheville, N. Car. 28804   po
Crops: corn
Livestock: 35-40 head cattle

Dr. John Lockard   1
28 Battery Park Ave., Box 944   pt
Asheville, N. Car. 28802   ao
Crops: potatoes, corn, beans

Clarence Chappell, Jr.   200
Belvidere,   ft
N. Car. 27919   ao
Crops: vegetables, oats, wheats
Livestock: hogs, beef cattle, chickens
Regular commercial channels

Mrs. Jack Sheridan   10
RD 1   pt
Sherwood Forest   ao
Brevard, N. Car. 28712

James Nuckles   20
Rt. 1, Box 814-A   pt
Colfax, N. Car. 27235   ao
Crops: garden variety
Livestock: pigs, chickens, rabbits

Michael & Rosemary Surles   12
Rt. 1   ft
Culberson, N. Car. 28903   ao
Crops: vegetables
Direct to retail store
Regular commercial channels

Vernie C. Grady   100
Box 166   ft
Dudley, N. Car. 28333
Crops: corn, tobacco, wheat, soybean, truck
crops, strawberries
Livestock: 26 hogs, 3 calves
Regular commercial channels

F. Sawyer Farms   99
Rt. 2, Box 141-147
Grifton, N. Car. 28530
Crops: sweet potatoes, corn, grain variety,
artichokes

Jacob Shepard   50
Rt. 4, Box 208   pt
Jacksonville, N. Car. 28540   po
Crops: tobacco and stock
Quantities: 9000 lbs. tobacco
Livestock: 40 head and 25 hogs
Wholesaler
Regular commercial channels

Robert C. York   75
910 10th St.   ft
Lillington, N. Car. 27546   po
Crops: tobacco, corn, soybeans, wheat
Livestock: 25 brood cows
Regular commercial channels

100

Hillcrest Gardens   5
Max Satterwhite   ft
Rt. 5   po
Oxford, N. Car. 27565
Crops: flowers & vegetables & fruits
Roadside stand

Franklin P. Cox   50
Rt. 1   pt
Randleman, N. Car. 27317   po
Crops: tobacco, strawberries
Quantities: 10,000 lbs. tobacco, 150 qts. strawberries
Livestock: hogs

NORTH DAKOTA

George Andreas   200
Box 156   ft
Belfield, N. Dak. 58622   ao
Crops: wheat, oats, barley
Quantities: about 1000 bu. each
Livestock: few milk cows, some range cattle, chickens
Regular commercial channels

Edward Duletsky   160
Belfield,   pt
N. Dak. 58622   ao
Crops: wheat, oats, potatoes
Quantities: 850 bu.
Regular commercial channels

Michael Beck   40
R.R. 2, North River Rd.   pt
Bismarck, N. Dak. 58501   ao
Crops: alfalfa, raspberries, onions
I sell my harvest to local customers.

E. J. Hilber   35
921 8th St. S.   pt
Fargo, N. Dak. 58102   ao
Crops: fruits and vegetables
Livestock: 10 head beef
Roadside stand

Gregor L. Helland   62
Rt. 1   pt
Stanley, N. Dak. 58784   ao
Crops: wheat, oats
Quantities: 200-1000 bu.
Livestock: cattle
Regular commercial channels

OHIO

Mr. & Mrs. Merle Kinser   45
Amanda,   pt
Ohio 43102   po
Crops: corn
Quantities: 45000 lb. corn
Livestock: 8 heifers and 3 calves
Regular commercial channels

Ocena Enders   67
Rt. 1   ft
Attica, Ohio 44807
Crops: soybeans, wheat, oats
Quantities: 2000 bu.
Regular commercial channels

Jonas Nisley   45
County Rd. #60   ft
Baltic, Ohio 43804   ao
Crops: sunflowers, vegetables, grass
Livestock: beef
Sell through our own store.

Dennis Rush   30
2609 Harold Rd.   pt
Batavia, Ohio 45103   po
Crops: wheat, vegetables
Livestock: poultry and horses
Regular commercial channels

Robert Raynow   60
RD 1   pt
Bolivar, Ohio 44612   po
Crops: feed corn, hay, vegetables
Quantities: 2000 bu. corn, 30 tons hay
Livestock: pigs, hogs
Regular commercial channels

Lawrence B. Greve   80
Rt. 1   pt
Botkins, Ohio 45306   ao
Crops: corn, beans, oats, wheat
Quantities: 300 wheat, 1000 bu. corn, 800 oats
Livestock: beef, chickens, rabbits, hogs
Wholesaler, direct to retail store

Paul Rupp   35
Rt. 3   pt
Bryan, Ohio 43506   ao
Crops: oats, soybeans, vegetables
Livestock: beef cattle, poultry, hogs, goats
Roadside stand
Regular commercial channels

Richard Ries   90
11845 Portage St.   ft
Canal Fulton, Ohio 44614   po
Crops: corn, hay, oats, wheat
Quantities: 45 acres corn
Livestock: beef cattle

Glenn Daubach   11
4470 Waynesburg Dr. S.E.   pt
Canton, Ohio 44707   ao
Crops: sweet corn
Quantities: 1½ acres corn
Livestock: 300 chickens

Dr. George Beck   75
6315 Beechmont   pt
Cincinnati, Ohio 45230
Crops: hay, corn
Quantities: 12 bu. corn
Livestock: beef herd
Regular commercial channels

Chadwick Christine, Jr.   3
8550 Keller Rd.   ao
Cincinnati, Ohio 45232
Crops: tomatoes, melons

Marvin Westrich   ¼
1144 Wellspring Dr.   pt
Cincinnati, Ohio 45231   ao
Crops: tomatoes, lettuce

Steven R. Loper   50
2793 Atwood Terr.   pt
Columbus, Ohio 43211   ao

Mike Seiler   100
166 E. Frambes   ft
Columbus, Ohio 43201

George Bugos   30
RD 1   pt
Cortland, Ohio 44410   ao
Crops: oats, corn
Quantities: 50 bushels oats per acre, 2 tons corn per acre
Regular commercial channels

Mrs. W. J. Teagle   5
548 W. Steeles Corners Rd.   pt
Cuyahoga Falls, Ohio 44223   ao
Crops: tomatoes and squash
Livestock: horses, goats, chickens, ducks, rabbits
Roadside stand

Daniel F. Rupp   30
Box 54   pt
Delta, Ohio 43515   po
Crops: corn, soybeans
Quantities: 10 acres corn, 20 acres soybeans
Livestock: 8 pigs
Regular commercial channels

Bill Schmidlin   200
RD 1   ft
Delta, Ohio 43515   po
Crops: corn, soybeans, wheat, small 10 acre truck patch
Roadside stand
Regular commercial channels

101

Richard Bailey 150
Box 247 ft
East Orwell, Ohio 44034 po
Crops: corn, oats, alfalfa, milk
Quantities: 4,000 bu. corn, 2,500 oats, 100 tons
alfalfa hay, 600,000 lbs. milk
Livestock: 80 head holstein cattle
Regular commercial channels

Forrest Caldwell
42228 Hilltop Dr. pt
Elyria, Ohio 44035 po
Crops: beans, tomatoes, potatoes

Dietrichs Organic Farm 45
Rt. 1 pt
Genoa, Ohio 43430 ao
Crops: green edible soybeans, corn, hay
Wholesaler, direct to retail

Theodore Motter 121
23830 Bradner Rd., RD 1 pt
Genoa, Ohio 43430 ao
Crops: wheat, rye, soybeans
Sell to some old customers.
Regular commercial channels

Col. Robert Curtis 144
R.R. 4 ft
Greenville, Ohio 45331 po
Crops: onions, tomatoes, corn, oats, cabbage,
beans
Quantities: 3, 50, 80, 12, 12, 1, 5 acres
Livestock: chickens, hogs, beef, goats
Regular commercial channels

Charles Spriggs 42½
RD 1 pt
Hanoverton, Ohio 44423 po
Crops: wheat, oats, eggs, beef
Livestock: cattle, chickens, goats, guineas
Regular commercial channels

Glenn Graber 400
13737 Duquette Ave. ft
Hartsville, Ohio 44632 po
Crops: potatoes, radishes, onions, endive,
escarole, parsley, lettuce, squash, pumpkins,
field corn, oats, rye
Quantities: 300-400 truckloads
Livestock: cattle
Regular commercial channels

Abe D. Schlabach 90
R. 1 ft
Holmesville, Ohio 44633 ao
Crops: hay, wheat, corn & oats
Quantities: 1050 bu. small grain, 1000 bu. corn
Livestock: horses, cows, hogs and chickens
Regular commercial channels

Chas. Zilich 25
RD 1 pt
Huntsburg, Ohio 44046 ao
Crops: beef, vegetables, fruit, maple syrup
Livestock: 4 herefords
Sell through relatives and friends.

James Terry 70
1817 Stumpville Rd., RD 2 pt
Jefferson, Ohio 44047 ao
Crops: legumes, 4th year of soil building
Wholesaler

L. C. Snavely 60-65
3516 Meadow Lane
Kettering, Ohio 45419
Interested in organic farming
Wholesaler

Benedict Treglia 33
3220 N. Pevee Rd., Rt. 7 pt
Lima, Ohio 45854 ao
Crops: tomatoes, lettuce, carrots, beets, squash,
peppers
Quantities: 50 bu.
Livestock: 12 cattle, 9 horses
I have my own store.

Robert Wickline 80
R.R. 2, Box 89 pt
Lynchburg, Ohio 45142 po
Crops: corn
Livestock: hogs and cattle
Regular commercial channels

Richard Chickletts 40
RD 3 pt
Mantua, Ohio 44255 po
Crops: oats, hay, large garden
Quantities: 60 bu. per acre
Livestock: chickens, pigs
Wholesaler
Regular commercial channels

Charles Molnar
1246 Summit Dr. pt
Mayfield Hgts., Ohio 44124 ao
Crops: fruits and vegetables—40 apples, 15
pears, 10 peach, 10 cherry, 25 plum, 5 cherry-
plum cross, 10 Juneberry, 50 sand cherry, 25
buffalo berry, 50 blueberry, 20 hazelnut, 5
hicans, 200 black walnut, 10 butternut hybrids,
6 English walnuts, 500 blackberry, 500 raspberries, 100 asparagus, 20 rhubarb, 100 elder-
berry, 50 currants.

William Scamahorn 160
6248 Kenyon Ct. pt
Mentor, Ohio 44060 po
Crops: wheat, corn, oats, etc.
Wholesaler
Regular commercial channels

J. F. Edgington 30
Edge-of-Sight Farm ft
Box 216, R.R. 1 ao
Millfield, Ohio 45761
Crops: soybeans, berries, misc. veg., nuts, eggs
Direct to retail store
Regular commercial channels

Howard Oliver 170
Rt. 2 pt
Mount Orab, Ohio 45154 po
Crops: corn, wheat, soybeans
Quantities: 4000 bu. corn, 1500 bu. wheat, 1600
bu. soybeans
Regular commercial channels

Harold G. Chamberlain 102
RD 2 pt
New Concord, Ohio 43762 ao
Crops: apples
Quantities: 6000 to 8000 bushels
Wholesaler, roadside stand, direct to retail
store
Regular commercial channels

The Way Inc. 100
Box 328 ft
New Knoxville, Ohio 45871 po
Crops: corn, beans, wheat
Regular commercial channels

Mr. & Mrs. Dayton E. 25
    Livingston pt
Rt. 10, E. ao
Oberlin, Ohio 44074
Crops: feed for cattle
Livestock: beef cattle, chickens
I sell direct to consumer.

Paul Siedel 1½
27353 Schady Rd. pt
Olmstead Falls, Ohio 44138 po
Crops: squash and corn
Livestock: chickens
Direct to retail store
Regular commercial channels

Keneil Blaho 185
8230 W. Rt. 718 ft
Pleasant Hill, Ohio 45359 po
Crops: hay, wheat, corn, soybeans
Livestock: cattle, chickens
Regular commercial channels

John A. Lautzenheiser 25
RR
Ridgeway, Ohio 43345
Plan to operate totally organic farm and to
grow vegetable crops to sell at the roadside
stand we plan to build.

James Robinett 42
RD 1, Box 138 pt
Rittman, Ohio 44270 po
Crops: hay, corn, truck gardening
Livestock: sheep, hogs and beef
Wholesaler
Regular commercial channels

102

John McClary 115
Rt. 2 pt
Sabina, Ohio 45169 po
Crops: corn and oats
Livestock: cattle and hogs, few chickens
Roadside stand

Ray V. Satterfield 25
Rt. 3 pt
Thornville, Ohio 43076 po
Crops: corn
Livestock: horses, cows, chickens
Roadside stand

Amos Brubaker 78
Box 158, RR 5 ft
Tiffin, Ohio 44883 ao
Crops: soybeans, wheat, oats
Quantities: 1200 bu. beans, 350 bu. wheat, 150
bu. oats
Livestock: 1 steer, 3 goats
Regular commercial channels

H. E. Zirger 135
R. 4 ft
Tiffin, Ohio 44883 ao
Crops: corn, beans, oats, wheat, apples
Quantities: 2000 bu., 90 bu., 1000 bu., 800 bu.,
20 bu.
Livestock: beef (10 head), lamb, chickens
Wholesaler
Regular commercial channels

Gilbert J. Calta 50
P.O. Box 321 pt
Valley City, Ohio 44280 po
Crops: hay and sweet corn
Livestock: sheep and cattle, goats
Roadside stand
Regular commercial channels

Dean Koch 31
R.R. 6 pt
Wapakoneta, Ohio 45895 po
Crops: tomatoes and sweet corn
Roadside stand and direct to retail store.

Richard P. Gainok 52
RD 1, Box 198 pt
Wellington, Ohio 44090
Crops: corn, alfalfa, timothy
Quantities: 15 acres corn, 12 acres timothy
Livestock: steers, sheep, chickens, ducks, &
horses
Roadside stand

Dale Scott 200
RD 1 ft
Willard, Ohio 44890 po
Crops: corn, oats, wheat, hay and soybeans
Livestock: feeder pigs, sheep and a few steers
Regular commercial channels

Indian Oaks Farm 30
181 E. Wilson Bridge Rd. pt
Worthington, Ohio 43085 ao
Crops: hay, potatoes and garlic
Quantities: 20 tons hay, 50 bu. potatoes, 20
bu. garlic
Wholesaler and direct to consumer

Mr. & Mrs. J. R. 40
Rhodehamel pt
Sperling Lane, R.R. 6 ao
Xenia, Ohio 45385
Crops: cattle, fruit, truck (small)
Livestock: cattle, rabbits, chickens

OKLAHOMA

Mrs. Robert Scott 320
Rt. 3
Antler, Okla. 74523
Plan to raise garden produce to sell as well as
for own use.
Livestock: 50 cows and 30 calves

Mr. & Mrs. D. W. Couch 50
1224 Seventh NW pt
Ardmore, Okla. 73401 po
Crops: sweet potatoes, peanuts

Iris P. Colwell 14
Box 25 ft
Balko, Okla. 73931 po
Crops: wheat
Quantities: 150 bu.
Livestock: sheep
Regular commercial channels

Lonnie W. Jourdan 5
P.O. Box 706 pt
Chickasha, Okla. 73018 po
Crops: vegetables and fruit
Livestock: horses, lambs, calves
Just started this year.

H. Roger Quimby 90
Rt. 1, Double Ranch pt
Collinsville, Okla. 74021 po
Crops: hay
Quantities: 50 tons
Livestock: hogs, cattle and quarterhorses
Regular commercial channels

Mrs. R. F. Holland 50
Rt. 1, Box 223 ft
Durant, Okla. 74701 po
Crops: peanuts, cotton, hay and some vege-
tables
Quantities: 1500 lbs. to 3000 lbs. peanuts, 2 to
3 bales cotton, 15 to 20 tons hay
Livestock: cattle, hogs and few chickens
Regular commercial channels

R. L. Zickefoose 320
1621 W. Bdwy. ft
Enid, Okla. 73301 ao
Crops: wheat and barley
Quantities: 5000 bu. of each
Regular commercial channels

Frank Ferchan, Jr. 1½
Box 224 ft
Gage, Okla. 73843 ao
Crops: rye pasture, vegetables
Livestock: registered jersey milk cows
Direct to retail store and to my own customers.

Hubert West 40
R.R. 1, Box 124 pt
Glencoe, Okla. 74032 po
Crops: wheat, rye, vetch
Quantities: 800 bu.
Livestock: cattle, goats
Regular commercial channels

Jim Zimmerman
Box 96 pt
Hailett, Okla. 74034 po
Crops: fruit and vegetables
Livestock: few cattle, chickens and 2 hogs
Sales are made from our home.

George Lowery 260
Rt. 2, Box 35 ft
Kingfisher, Okla. 73750 po
Crops: wheat, alfalfa
Quantities: 2000 bu. wheat, 555 tons alfalfa
Livestock: 50 head steer
Direct to retail store
Regular commercial channels

Don G. Morris 100
Box 246 ft
Sapulpa, Okla. 74066 ao
Crops: hay feed, fruit and berries
Quantities: 5 acres fruit & berries
Livestock: horses and cattle

Charles Hallmark 60
Rt. 2, Box 83 pt
Sulphur, Okla. 73086 po
Crops: vetch and rye
Quantities: 33.7 acres of each
Livestock: cattle, hogs, rabbits
Regular commercial channels

Larry Warden 1
Rt. 1, Box 95 pt
Tishomingo, Okla. 73460 po
Crops: small veg. garden for family
Livestock: Holstein heifers for dairy use

Paul Garfield 100
1320 E. 19th St. ft
Tulsa, Okla. 74120 ao

Carl L. Barnes   180
Turpin,   pt
Okla. 73950   po
Crops: wheat, sorghum, corn
Quantities: 750 bu. wheat, 1200 bu. sorghum,
100-500 bu. whole corn
Sell direct to elevators.

OREGON

G. S. Hudelson   2
Rt. 1, Box 830   pt
Boring, Ore. 97009   ao
Crops: garlic, blueberries
Quantities: 100 lb. garlic
Direct to retail store
Regular commercial channels

Betty Quimby   31
Rt. 1, Box 188   ft
Cornelius, Ore. 97113   ao
Crops: garden vegetables
Livestock: horses, goats, rabbits, chickens
Regular commercial channels

Norman Forsberg   48
Rt. 3, Box 1   pt
Dallas, Ore. 97338   ao
Crops: fruit
Livestock: sheep, goats and beef cattle
Direct retail sales and cannery sales

Henry Odegard   40
Rt. 1, Box 1038   pt
Eugene, Ore. 97402   ao
Crops: prunes, variety fruits and vegetables
Sell my harvest to farmers market.

Edward Baker   25
Rt. 3, Box 207   pt
Hillsboro, Ore. 97123   ao
Crops: vegetables & 5 acres prunes
Livestock: rabbits, chickens and horses
Roadside stand

Mrs. Elmer Panko   2
Rt. 3, Box 404B   ft
Hillsboro, Ore. 97123   ao
Crops: variety of vegetables
Livestock: 5 beef cattle, 5 horses, 50 chickens
& ducks
Direct to retail store and customers to farm.

Samuel R. Powell   10
Rt. 2, Box 211-W   pt
Hillsboro, Ore. 97123   ao
Crops: vegetables
Livestock: cattle, horses, hogs
I sell my harvest through word of mouth.

V. M. Caley   70
2427 Corvallis St.   pt
Klamath Falls, Ore. 97601   po
Crops: alfalfa, rye, hay
Quantities: 50 tons
Livestock: 40 beef cattle
Regular commercial channels

Sunray Orchards   55
John M. Herman   ft
Rt. 1, Box 299   ao
Myrtle Creek, Ore. 97457
Crops: prunes
Quantities: 100,000 lbs.
Wholesaler, direct to retail store and mail-
order.

Luke Brooks   570
Rt. 2, Box 125A   ft
Pendleton, Ore. 97801   po
Crops: wheat, barley, elephant garlic, arti-
choke, zucchini, squash
Quantities: 900 lbs. garlic
Livestock: hereford & holstein
Wholesaler and direct to retail store
Regular commercial channels

Russell Shepeard   22
6902 Fruitland Rd. N.E.   pt
Salem, Ore. 97301   po
Crops: garlic
Livestock: 20 head cattle
Roadside stand

R. D. Stagner   30
Shallon Farm   ft
971 Avenue E   ao
Seaside, Ore. 97138
Crops: globe artichokes
Quantities: 6,000-8,000 lbs. per acre

York & Joanna Rentsch   40
Route 1, Box 175   pt
Sheridan, Ore. 97378   ao
Crops: alfalfa, grains, pasture, new orchard
Size: all small
Livestock: dairy, beef, poultry
Regular commercial channels

George Weideman   50
Rt. 3, Box 75   pt
Springfield, Ore. 97477   po
Crops: beef cattle
Livestock: 15 head cattle, pigs, chickens

Mr. & Mrs. Mike Michaels   5-30
R.R. 1, Box 93   ft
Vale, Ore. 97918   pt
Crops: vegetables, grains, berries and fruit trees
Livestock: goats, chickens, geese, sheep, yaks,
peacocks
Sell harvest through wholesaler if available
roadside stand, retail store if available and
mail order.

Mrs. Pearl Terry   150
Rt. 1, Box 168   ft
T.S.T. Ranch   po
Wallowa, Ore. 97885
Crops: wheat
Quantities: 6000 ton
Livestock: 50 cattle
Regular commercial channels

Green Parrot Goat Farm   6
Rt. 1, Box 255   ft
Willamina, Ore. 97396   ao
Crops: truck garden
Roadside stand

John Piumarta   50
Rt. 1, Box 368   ft
Yoncalla, Ore. 97499   ao
Livestock: 30 head yearly

PENNSYLVANIA

Andrew Gallant   60
RFD 2   ft
Albion, Pa. 16401   ao
Crops: garlic
Quantities: 30,000 lb.
Livestock: hogs, chickens
Sell harvest through wholesaler, roadside stand,
and direct to retail store.
Regular commercial channels

Full Circle Farm   65
RD #1
Beavertown, Pa. 17813
Crops: apples, peaches, pears

Charles Henry   50
Box 127, RD 4   pt
Bedford, Pa. 15522
Just bought farm

Jesse M. Symons   100
RD 4, Box 37   pt
Belle Vernon, Pa. 15012   po
Crops: corn, hay
Quantities: 5,000 bu. corn, 10,000 bales hay
Regular commercial channels

Wilson Marts   20
RD 1   pt
Berlin, Pa. 15530   ao
Crops: potatoes, strawberries, corn
Quantities: 150 bu., 500 qts., 60 doz. corn
I sell my harvest door to door.

Richard Mogel   40
RD 1   ft
Bernville, Pa. 19506   ao
Crops: corn, hay, vegetables
Livestock: Jersey cows, heifers

Ken Deitch 30
RD 1 pt
Boiling Springs, Pa. 17007 po
Crops: strawberries, raspberries, blueberries
Sell harvest direct to retail store.

Donald Miller 15
RD 1 pt
Brogue, Pa. 17309 ao
Crops: corn, melons
Livestock: dairy goats, sheep
Roadside stand

Cyrus Swope 60
Rt. 1 ft
Brogue, Pa. 17309 ao
Crops: wheat, corn, tomatoes, potatoes, berries
Quantities: 40 ton tomatoes, 1000 qt. blue-
berries, 200 potatoes
Livestock: beef cattle
Sell harvest to Farmers Market in York.
Regular commercial channels

Robert W. Tirrell 30
RD 1 ft
Cambridge Springs, Pa. 16403 ao
Crops: hay, maple syrup and milk
Quantities 1000 gals. per year
Livestock: registered Jersey cattle
Roadside stand
Regular commercial channels

Wilbur Wolf, Jr. 149
RD 4 pt
Carlisle, Pa. 17013 po
Crops: beef cattle, eggs
Quantities: 10-15 steers/year — 10 doz. eggs/
week
Livestock: 14 cows and heifers aiming for a
herd of 25
Sell direct to individuals through radio ads and
health food stores.

Mrs. Glenn Ford 25
RD 1 ao
Carlton, Pa. 16311
Crops: oats, garlic
Quantities: 480 bu. oats, 3 bu. garlic
Livestock: 20 head

Mrs. William Malone 57
Box 290, Rt. 1 pt
Carmichaels, Pa. 15320 po
Crops: vegetables and grain for stock
Livestock: chickens, pig, bees
Roadside stand

J. R. Woodruff 80
RD 1 pt
Chester Springs, Pa. 19425

L. M. Merk 185
Hunters Hill, RD 2 pt
Clearville, Pa. 15535 po
Crops: grains, hay, vegetables, fruit, corn,
wheat
Livestock: chickens, sheep, beef, pheasant,
quail
Roadside stand
Regular commercial channels

Norman & Marjorie Aamodt 276
Snowhill Farm pt
Coatesville, Pa. 19320 ao
Crops: beef
Livestock: Angus beef cattle

Charles Fulmer 65
Cogan Station, Rt. 1 ft
Pa. 17728 po
Crops: cereal grains, hay, strawberries
Livestock: dairy and capons
Regular commercial channels

Richard Dippold 100
RD 3 pt
Columbia Cross Roads, Pa. ao
  16914
Crops: beef
Sell to private buyers

Edward Sewalk 2
45 Ramsey St. pt
Conemaugh, Pa. 15909 po
Crops: potatoes, corn, beans, cabbage, tomatoes

Albert D. Knoblach 40
RD 3 pt
Confluence, Pa. 15424 po
Crops: hay & truck crops
Quantities: 70 tons hay
Livestock: 130 sheep, 5 cows
Roadside stand

Walter Loncosky 80
Rt. 3 ft
Danville, Pa. 17821
Crops: feed grains, vegetables, milk
Livestock: dairy goats, cows, sheep, poultry
Regular commercial channels

Mr. & Mrs. Leroy Nailor, Jr. 45
Ridge Rd., RD 3 pt
Dillsburg, Pa. 17019 po
Crops: complete line veg. and field corn, pota-
toes, tomatoes, beans, sugar peas and onions.
Livestock: capon geese, guineas, laying hens,
50 Hereford beef cattle, hogs
I sell harvest direct from farm.

Wm. J. Henry 20
158 Glen Drive pt
Doylestown, Pa. 18901
Crops: apples
Quantities: 8000 bu.

Jeffrey Chase 25
RD 1 pt
Dushore, Pa. 18614 ao
Crops: berries, tree fruit, corn, sunflowers
Livestock: chickens, ducks milk cows

Col. Frank Reidelbach 5
902 W. Highland Avenue pt
Ebensburg, Pa. 15931 ao
Crops: corn and green beans

Keith Thompson 32
1424 S. Pleasant Dr. pt
Feasterville, Pa. 19047 po
Crops: soybean, herbs (comfrey)
Quantities: very limited
Livestock: goats
Roadside stand

Enos Burkholder 100
Rt. 2 ft
Fleetwood, Pa. 19522 po
Crops: corn, grain, hay, plus vegetables, mostly
potatoes, soybeans, salad greens, peas, beans.
Livestock: 75 head cattle for beef
Wholesaler and sell some at house
Regular commercial channels

Deep Dale Farm 70
Pfahl & Sons ft
Box 129, RD 1 po
Forest City, Pa. 18421
Crops: dairy, 100 ft. x 100 ft. organic garden
Livestock: 50 heads
Roadside stand

R. H. Goshorn 150
329 Skippack Pike pt
Ft. Washington, Pa. 19034 po
Crops: corn, soybeans and wheat
Regular commercial channels

Richard Dallas 27
Dublin Pike pt
Fountainville, Pa. 18923 ao
Crops: beef, small grain, hay
Quantities: 4 steers
Livestock: beef, pigs, broilers

Noll's Nursery 3
RD 1, Box 152 pt
Georgetown, Pa. 15043 po
Crops: artichokes, tomatoes, beans, cucumbers,
green onions
Sell harvest to individual customers.

Stanley Beehler 75
Mt. View Farm, RD 1 ft
Gilbertsville, Pa. 19525 po
Crops: hay and dairy products
Livestock: 11 cows and 9 heifers
Wholesaler
Regular commercial channels

Mrs. Charles Kuzmen 25
RD 4, Box 193 pt
Greensburg, Pa. 15601 po
Crops: grain
Livestock: few cows, pigs and chickens

William Lee Nice 30
Box 475, Stover Rd. pt
Harleysville, Pa. 19438 po
Crops: peaches, corn, grain
Quantities: 13 tons, 25 tons, 1 ton
Livestock: 1 cow, 6 steers, 30 laying hens
Roadside stand

Edwin McKeever 27
5851 Union Deposit Rd. pt
Harrisburg, Pa. 17111 po
Crops: grain, fruit and truck
Market harvest through local media.

Herb Green 30
Star Route Seelyville ft
Honesdale, Pa. 18431 ao
Livestock: horses, chickens, asst. small animals

William Albright 81
16 W. Main St. pt
Hummelstown, Pa. 17036 po
Crops: sweet corn, strawberries
Livestock: steers
Roadside stand

Mrs. Lester Walters 30
Rt. 4 pt
Hummelstown, Pa. 17036 ao
Planted in grass except 1 acre garden
Direct to retail store
Market to established customers and friends.

Willis A. Kinsey 70
RD 1 pt
Julian, Pa. 16844 po
Livestock: beef, horses, chickens
Regular commercial channels

J. Arnold Voehringer 30
RD 2 pt
Kempton, Pa. 19529 po
Crops: corn, oats, clover, timothy, alfalfa
Quantity: 15 tons hay, 500 bu. corn
Livestock: 8 steers, 1 heifer, 2 pigs

Gordon Gochenaur, Inc. 40
123 Locust St. ft
Lancaster, Pa. 17602
Crops: corn, wheat, pasture
Livestock: hogs and cows

Kenneth Clouser 130
Madisonburg, ft
Pa. 16852 po
Crops: corn, oats, hay
Quantities: 3000 bu. corn, 1500 bu. oats, 90 tons hay
Livestock: 50 head dairy cows, chickens, horses, calves
Regular commercial channels

Will A. Seegers 1
Malvern, ft
Pa. 19355 ao
Crops: garden vegetables—beets, beans, cabbage, carrots, swiss chard, parsley, zucchini, tomatoes, rhubarb
Direct to customers who call

Charlotte Waldron 5½
609 Lancaster Pike ft
Malvern, Pa. 19355 ao
Crops: honey, eggs, poultry, sweet corn, asparagus, carrots, tomatoes
Quantities: 1½ tons, 1100 doz. 60-70 dressed poultry, 2-3 acres 200 lbs. 10 bu. 200 lbs.
Livestock: laying hens, roasting chickens, geese, ducks, sheep, goat, rabbits, pigeons
Direct to retail store and people who come to house.

Kenneth Yoder 40
RD 4, Box 252 pt
Meadville, Pa. 16335 po
Crops: hay, oats, corn, berries
Quantities: 20 acres hay, 6 acres corn, 6 acres oats
Livestock: 1 cow, 3 heifers, 3 beef cattle, turkeys, chickens, pony

Cockley's Rocky Ridge 5
Asparagus Farm, RD 5 pt
Mechanicsburg, Pa. 17055 ao
Crops: asparagus
Wholesaler

Marvin Rhodes 45
Millmont, ft
Pa. 17845 ao
Crops: tomatoes, beans, turnips, corn, hay, wheat, rye
Quantities: 1000 doz. corn, 50 tons tomatoes, 2 tons beans, 300 bu. wheat, 300 bu. rye, 1 ton turnips, 4 acres sweet corn
Livestock: cattle
Sell mostly to Walnut Acres

Robert Fleming 121
North Hills ao
Milton, Pa. 17847
Crops: oats, wheat, soybeans
Livestock: 13 head
Regular commercial channels

Mr. & Mrs. Courtland 30
  Birchard ft
RD 5 po
Montrose, Pa. 18801
Crops: hay, oats, corn, vegetables
Livestock: 23 head beef cattle, 2 horses
Plan on selling to campers.

Lawrence P. Delp 100
218 N. 25th St. pt
Mt. Penn, Pa. 19606 ao
Crops: alfalfa, sorghum
Quantities: variable
Livestock: pheasant, quails

William Robbins 6
RD 2 pt
Muncy, Pa. 17756 ao
Crops: vegetables, poultry
Livestock: dairy goats, sheep, rabbits
Direct to retail store

Arthur Shutt, Jr. 150
RD 3 ft
Muncy, Pa. 17756 po
Crops: corn, oats, wheat, hay
Quantities: 30 acres, 30 acres, 15 acres, 30 acres
Livestock: dairy and swine

Aaron Bersole 20
RD 3, Box 355 ft
Myerstown, Pa. 17067 ao
Crops: vegetables, strawberries, red raspberries, cherries
Livestock: 4 angus steers, 100 chickens, veal calves
Direct to retail store and any other way we can.

Enos H. Hess 86
RD 1 ft
Myerstown, Pa. 17067 ao
Crops: white potatoes, sweet potatoes, beets, carrots, onions
Livestock: beef
Direct to retail store and mail order and pick up at farm.

Clarence A. Derr 100
RD 1 ft
New Bloomfield, Pa. 17068 po
Crops: blueberries, strawberries, raspberries
Regular commercial channels

Barkley Baldwin 4
Pineville Rd., RD 2 pt
New Hope, Pa. 18938 ao
Crops: strawberries, rhubarb, timothy, and personal veg. garden
Livestock: chickens, horse
Sell harvest to private families desiring non-polluted food.

Rudy D. Byler 80
RD 1 ft
New Wilmington, Pa. 16142 ao
Crops: wheat, oats, corn, potatoes, hay
Quantities: 18 acres each except potatoes
Livestock: cows, hogs, horses, chickens
Feed most harvest to livestock

106

Vincent N. Yutko   27
RD 1   pt
Orwigsburg, Pa. 17961   po
Crops: vegetables, corn, hay
Livestock: 4 hogs, 3 heifers
Roadside stand

L. C. Kemp   7
RD 3, Box 298   pt
Oxford, Pa. 19363   ao
Crops: vegetables, tree fruits and brambles
I sell my harvest direct.

Joseph Lapp   70
Rt. 1   ft
Paradise, Pa. 17562   ao
Crops: wheat, corn
Livestock: dairy beef
Regular commercial channels

Wilbert Walker   17
Walkers Organic Acres   pt
Main & Callowhill Sts.   ao
Perkasie, Pa. 18944
Crops: sweet corn, cantaloupes, watermelon, tomatoes, peppers, squash, pumpkins, turnips.
Wholesaler, roadside stand, direct to retail store

Mrs. Richard Rimbach   28
8550 Babcock Blvd.
Pittsburgh, Pa. 15237
Roadside stand

Scott Frymoyer   40
RD 1   pt
Port Trevorton, Pa. 17864   po
Crops: corn, oats, wheat, hay
Quantities: 168 bu. wheat, 270 oats, 600 corn
Livestock: veal calves, steers, cattle
Wholesaler
Regular commercial channels

D. L. Geinett   23
Richfield,   ft
Pa. 17086   ao
Crops: corn, oats, hay pasture
Livestock: beef and lamb, geese
Wholesaler
Regular commercial channels

J. D. Roboski, Sr.   20
RD 1   pt
Roulette, Pa. 16746   ao
Crops: garden vegetables
Quantities: 2 acres
Roadside stand

Forrest W. Strawbridge   10
197 Miller Rd.   pt
Sinking Spring, Pa. 19608   ao
Crops: strawberries, cantaloupes, sweet corn, cucumbers, tomatoes
Livestock: 1 pony
Sell my harvest at my home and door to door.

Nelson Brenneman   83
R.D. 1, Box 397   ft
Spring Grove, Pa. 17362   ao
Crops: alfalfa, field corn, vegetables
Quantities: 500 tons alfalfa, 1200 bu. corn
Livestock: dairy cows, chickens, rabbits
Roadside stand, regular commercial channels

H. R. Lefever   3
RD 1, Box 457   pt
Spring Grove, Pa. 17362   ao
Crops: misc. vegetables & fruits, soybeans
We have our own health food store located on our farm.

Muller Stock Farm   200
Lucas Muller   ft
RD 2, Box 204   ao
Stroudsburg, Pa. 18360
Crops: oats, alfalfa, corn, wheat
Livestock: beef, Angus & Charolais
Regular commercial channels

Mrs. Harold Smith, Jr.   20
RD 2
Summerville, Pa. 15864
Crops: vegetables
Livestock: 1 cow, veal calves

Irvin G. Zimmerman   55
513 Cedar Lane   pt
Swarthmore, Pa. 19081   ao
Crops: beef, apples, grapes (wine)
Quantities: 1,000 bu. apples & 100 gals. wine
Livestock: 10 angus, 500 chickens
Roadside stand

John Williams   25
RD 4, Box 189A   pt
Tarentum, Pa. 15084   po
Crops: potatoes, tomatoes, corn
Livestock: 5 beef cattle

Eugene Gettig   40
RD 3   pt
Titusville, Pa. 16359   ao
Crops: oats, barley, corn, buckwheat
Quantities: 300 bu. oats, 100 bu. barley, 500 bu. corn
Livestock: cattle, chickens, horses, rabbits
Wholesaler and auctions, etc.
Regular commercial channels

Mr. Richard Eaton   90
RD 2, Box 160   pt
Troy, Pa. 16947   po
Crops: corn, beet
Quantities: 800 bu.
Livestock: 20 holstein
Regular commercial channels

Norman Steck   195
RD 1   ft
Turbotville, Pa. 17772   po
Crops: corn, oats, wheat, hay
Livestock: hogs and beef
Regular commercial channels

Paul F. Hagenbuch   90
RD 1   pt
Watsontown, Pa. 17777   po
Crops: tomatoes and sweet corn
Quantities: 3 acres
Roadside stand

Robert Himes   30
RD 2   pt
Williamsburg, Pa. 16693   po
Crops: corn, oats and hay
Livestock: beef and hogs

Ray A. Machamer   6
319 Vine St.   pt
Williamstown, Pa. 17098   ao
Crops: potatoes, corn, tomatoes and onions
Livestock: 1 bull and 2 pigs
I sell my harvest to anyone.

RHODE ISLAND

Rita C. Broady   3
Rt. 2, Box A6   pt
Foster, R.I. 02825   ao
Crops: tomatoes, beans, squash, greens, etc.
2 acres cranberry bog
Livestock: 1 horse, 1 pony, 1 calf, 1 milk cow
Roadside stand, direct to retail store and co-op groups
Market harvest anyway I can.

R. I. Moore   50
706 Main St.   pt
Wickford, R.I. 02852   ao
Crops: all vegetables
Livestock: pigs, beef cattle, chickens

SOUTH CAROLINA

L. M. Gillespie   25
Rt. 3, Box 450   pt
Easley, S. Car. 29640   po
Crops: general
Livestock: some

Gerry Keys   30
Rt. 6, Box 245   pt
Gaffney, S. Car. 29340   ao
Crops: all garden crops, fruits and berries
Livestock: chickens, soon have beef and hogs

T. C. Wilson   35
RFD 1   ao
Williamston, S. Car. 29697
Crops: strawberries
Livestock: 6 to 12 head cattle
Roadside stand

SOUTH DAKOTA

C. M. Rhody   200
Bemis,   ft
S. Dak. 57215   ao
Crops: oats, corn, wheat, flax
Livestock: Hereford cattle, chickens and goats
Regular commercial channels

Mrs. Charles W. Mader   100
P.O. Box 86   ft
Eagle Butte, S. Dak. 57625   ao
Crops: lamb, wheat, oats
Quantities: 18,000 lbs. lamb, 400 bu. wheat,
3,000 bu. oats
Livestock: 40 head cattle, 6 calves, lamb and
poultry
Direct to retail store
Regular commercial channels

Mrs. Helen Anderson   700
Box 602   ao
Martin, S. Dak. 57551
Crops: wheat, barley, oats, rye
Regular commercial channels

Glen Houdersheldt   44
1413 E. 5th   ft
Mitchell, S. Dak. 57301   ao
Crops: wheat, corn, eggs, beef, sunflower seeds,
onions, carrots
Livestock: chickens, cattle
Direct to retail store
Regular commercial channels

Gary Schmeichel
Parker,   ao
S. Dak. 57053
Crops: grain, corn, oats, wheat, apples, plums
and vegetables
Livestock: beef cattle
Regular commercial channels
Sell direct to retail store

TENNESSEE

Mr. & Mrs. Ray Winfrey   40
R.R. 1   ft
Bon Aqua, Tenn. 37025   ao
Crops: mixed truck garden, corn
Livestock: Guernsey beef cattle, chickens
Direct to retail store and city friends

Carl Gorodetzky   17
Rt. 1, Box 305   pt
Cookeville, Tenn. 38501   ao
Crops: grass, asparagus, grapes, peanuts
Just starting

Walter Gryctko   1
Rt. 6   pt
Crossville, Tenn. 38555   ao
Crops: garden vegetables and blackberries
Livestock: 4 ponies
My harvest for home use.

Donald Forbes   20
Rt. 3, Box 210   ft
Dayton, Tenn. 37321   po
Crops: tomatoes, corn, soybeans, etc.
Regular commercial channels

James E. Harris   125
RFD 4   pt
Dyersburg, Tenn. 38024
Crops: cotton, soybeans, grass
Livestock: poultry

Bill Idol   30
215 Lynn Garden Dr.   ft
Kingsport, Tenn. 37660   po
Crops: tomatoes, onions, lettuce, cucumbers,
corn, radishes, okra
Livestock: chicken, cattle

William Carpenter
Rt. 1, Clinton Hwy.
Knoxville, Tenn. 37912
Plan to farm soon and will raise grain, fruit,
beans.

Mrs. E. D. Shipley
Stonewall Farm
R.R. 17   ao
Knoxville, Tenn. 37921
Crops: strawberries, raspberries, garden crops
Livestock: black Angus, 9 cows, 1 bull
Direct to retail store
Regular commercial channels

Mrs. Claude Arnold   1
Rt. 6, Stewarts Ferry Pk.   po
Lebanon, Tenn. 37087
Crops: vegetables, black walnuts
Sell small amount to friends.

Joseph Fry   4
Rt. 5   pt
Powell, Tenn. 37849   ao
Quantities: 15 bu. chestnuts
Sell my harvest through mail order business.

W. H. Netherton   64
305 Forrest St.
Watertown, Tenn. 37184
At present do not have crops or livestock, but
will make a small start next spring.

TEXAS

George Powell   20
2035 Dahlia
Amarillo, Texas 79107
Intend to start gardening next year with fruit,
melons.

O. Newell   1
2100 Metcalfe Rd.   pt
Austin, Texas 78741   ao
Crops: okra, tomatoes, cantaloupes
Quantities: 50 bushels
Livestock: dog, cat
Wholesaler and individuals, friends
Harvest it myself.

Mrs. P. T. Eichelberger   73
1601 Southwood   pt
Baytown, Texas 77520   po
Crops: 300 peach trees
Livestock: 5 horses

Donald H. Oldmixon   25
Rt. 1, Box 82   pt
Brownsville, Texas 78520   po
Crops: cattle and garden crops
Livestock: 20 calves/year
Wholesaler
Market through individual customers

Marcel Walker   145
Rt.   ft
Buckholts, Texas 76518   ao
Crops: corn, peas, milo, okra, squash
Quantities: 15,000 lbs.
Livestock: 16 cows, 16 calves and 40 hens
Regular commercial channels

E. B. Cartwright   20
Cartwright Groves   ft
Box 331   ao
Carrizo Springs, Texas 78834
Crops: citrus
Quantities: 1200 bu. average yield
Direct to retail store and consumers

D. J. Flournoy   16
P.O. Box 237
Castonville, Texas 78009
Starting to farm in Spring 1971
Leased as pasture
Livestock: leased hogs and yearling

Vernon Roper   26
10507 W. Lawn   pt
Dallas, Texas 75229   po
Crops: beef and table vegetables
Livestock: black Angus beef
Wholesaler
Regular commercial channels

A. B. Weatherread 66
2346 Freelandway
Dallas, Texas 75228
Crops: hay and Johnson grass
Plan to start on organic method about 5 acres
and increase in size each year.

Doran & Linda Williams 20-30
Rt. 4, Box 248A   pt
Elgin, Texas 78621   ao
Crops: vegetables, sweet potatoes, squash,
beans, etc.
Livestock: milk goats, ½ doz. beef cattle
Direct to retail store and natural food restaurants.

Melvin Scherer 35
1800 Twilight Dr. S.   ft
Ft. Worth, Texas 76116   po
Crops: citrus & vegetables
Regular commercial channels

J. A. Weaver 1
3010 Elm Park Dr.   pt
Fort Worth, Texas 76118   ao
Crops: vegetables and some fruit

Timothy Carlson 150
Rt. 2, Box 178   ft
Granger, Texas 76530   po
Crops: cotton and maize
Quantities: 2/3 bale per acre and 3500 lbs.
per acre.
Livestock: cattle, pigs, and sheep
Wholesaler
Regular commercial channels

Harold S. Baer 15
Rt. 1, Wilson Rd.   pt
Harlingen, Texas 78550   ao
Crops: yams, squash, tomatoes

Leland Davis 30
Star Route   pt
Henrietta, Texas 76365   po
Crops: grain and garden
Livestock: cattle

J. Frank Ford 1820
Arrowhead Mills, Inc.   ft
P.O. Box 866   ao
Hereford, Texas 79045
Crops: wheat and rye
Quantities: 25,000 bu. wheat, 5,000 bu. rye
Livestock: cattle

H. M. McKean 75
7714 Brykerwoods   pt
Houston, Texas 77055   po
Crops: hay
Quantities: 40 tons/yr.
Livestock: cattle, pigs
Regular commercial channels

Gerald A. Knapp 50
1910 Normandy Dr.   pt
Irving, Texas 75060   ao
Crops: pecans
Quantities: 3,000 within 10-12 yrs.

Burns Tilton 60
Box 1431
Jacksonville, Texas 75766
Crops: cover
Use harvest for building up greenery.

Claud Senn 490
Box 14   ft
Jayton, Texas 79528   ao
Crops: cotton, milo, cattle
Quantities: 75 bales cotton
Livestock: 25 cows
Regular commercial channels

Lonnie Moore 40
Rt. 1, Box 296
Leandep, Texas 78641
Crops: garden vegetables
Livestock: chickens, turkeys, cattle, sheep,
guineas

Gunard Friberg 30
Rt. 1   ft
Leesburg, Texas 75451   ao
Crops: truck-vegetables
Quantities: hundreds of bushels
Livestock: cattle
Sell harvest to farmers market.
Regular commercial channels

Mrs. F. O. Teten 2
Rt. 1, Box 283   ft
Liberty, Texas 77575   ao
Crops: vegetables of all kinds, pecans, persimmon, plums and other fruits.
Livestock: 30 head of cows and young calves

Mrs. Margret Wright 27
Wright's Organic Farm &   pt
 Bake Shop   ao
Rt. 1, Box 260
Mesquite, Texas 75149
Crops: In-season vegetable garden
Quantities: 150 bushels mixed vegetables
Livestock: 8 beef cows, 1 milk cow, 150 chickens, 50 ducks
Direct to retail store and customers who come
to farm.

Jefferson Organic Orchards 15
Rt. 1, Box 224   ft
Mission, Texas 78572   ao
Crops: citrus fruit
Quantities: 15 tons

Yancey L. Russell 15
Rt. 1, Box 92   ft
Natalia, Texas 78059   ao
Crops: fruits and pecans
Quantities: small
Livestock: 12 sheep, 4 head Angus cattle
Sell harvest to More Natural Food Store

Mrs. Ruby Louis 50 x 100 ft.
P.O. Box 162   ft
Odessa, Texas 79760   ao
Crops: vegetables
Quantities: 280 lbs. so far this spring
Sell direct to customers

Terry Don Morrison 160
Box 282   ft
Quitaque, Texas 79255   po
Crops: cotton, maize
Livestock: 30 feeder steers or heifers
Regular commercial channels

J. V. Edmonds 85
1209 Holly Dr.   pt
Richardson, Texas 75080   ao
Crops: wheat, pecans, apples
Wholesaler

Malcolm Beck 100
Rt. 13, Box 210   pt
San Antonio, Texas 78218   ao
Crops: clover seed, tomatoes, potatoes, okra
Quantities: 9000 lbs., 4000 lbs., 4000 lbs., 1000
lbs.
Livestock: 12 head of cattle, 60 hens
Roadside stand, direct to retail store and on
farm.

J. R. Collins, Jr. 45
Box 10126   pt
San Antonio, Texas 78210   ao
Crops: tomato, cucumber, squash, peas
Livestock: stocker calves and sheep

G. E. Jungjohan 80
Box 351   pt
Sanger, Texas 76266   ao
Crop: hay at present
Livestock: cattle

Mrs. Betty Pettit 75
Rt. 1, Box 21A   ft
Schulenburg, Texas 78956   po
No crops yet
Livestock: cattle, horses, sheep

Robert Moore 640
Box 291   ft
Shamrock, Texas 79079   ao
Crops: corn, milo grain, sorghum, rye, wheat,
mammoth sunflowers, millet
Quantities: 3500 lbs. per acre
Regular commercial channels

Chas E. Knandel 30
Smiley, ft
Texas 78159 po
Crops: hay, oats, clover
Quantities: 10 tons
Livestock: 25 cows & calves
Regular commercial channels

Jo Murphy & 40
Judy Thompson pt
R.R. 2 ao
Sunset, Texas 76270
Livestock: hogs

Jim Miller 31
Rt. 3, Box 334-E ft
Waco, Texas 76708 po
Crops: fruit, corn, peas
Quantities: 1800 bu. fruit
Livestock: pigs, chickens
Roadside stand

Running Springs Fruit & 160
Hog Ranch ft
Wellington, ao
Texas 79095
Crops: apples, alfalfa, cherries, berries, peaches, processed pork
Livestock: 50 sows, 500 pigs, 500 hogs
Wholesale, direct to retail store and people come.
Regular commercial channels

UTAH

J. W. Simkins 165
P.O. Box 36 ft
Enterprise, Utah 84725 po
Crops: hay, barley, lambs and beef
Quantities: 450 tons alfalfa
Livestock: lamb, beef
Regular commercial channels

Louis W. Larson irrigated—140
Garland, dry—160
Utah 84312 ao
Crops: hay, grain
Quantities: 300 lambs, 30 fat cattle
Livestock: sheep and cattle
Regular commercial channels

Douglas G. Hooper 20
Box 12 pt
Gusher, Utah 84030 ao
Crops: vegetables, fruit, alfalfa, wild asparagus
Livestock: calves, lambs, horses
Wholesaler, roadside stand

A. Keith Barben 700
P.O. Box 65 ft
Marysvale, Utah 84750 ao
Crops: hay and grain
Quantities: 2000 tons of hay, 100 tons of grain
Livestock: cattle and sheep
Regular commercial channels

Mr. & Mrs. Jesse Anderson 50
326 S. 200 E. pt
Salina, Utah 84654 po
Crops: alfalfa
Quantities: 5,000 bails

Intermountain Distributors
508 West Third N.
Salt Lake City, Utah 84116

Tom Gross 30
Box 86 pt
Sterling, Utah 84665 ao
Crops: potatoes, produce and eggs
Livestock: dairy, turkeys, chickens, beef, hogs
Regular commercial channels

VERMONT

Mrs. Jeremy Freeman 10
Jacksonville Stage pt
Brattleboro, Vt. 05301 po
Crop: hay, raise own vegetables on 100% organic basis
Livestock: angus cattle & horses
We sell no produce at moment, interest in country store and plan to add a natural foods annex.

Mrs. Jean Arrowsmith ½
RD 1 pt
Bristol, Vt. 05443 ao
Crops: vegetables and soft fruits

Mrs. Bebe Wicker 35
Charlotte, Vt. 05445 ft
Crops: raspberries, flint corn ao
Livestock: 1 cow, 1 horse, 12 hens

Alan T. Boutiller 100
RFD ao
Hinesburg, Vt. 05461
Will start to farm next year.

Stuart C. Harnish 50
Box 127B, RD 1 ft
Jericho, Vt. 05465
Crops: vegetables
Roadside stand and direct to retail store

Glendon E. Wilder 40
N. Clarendon, ao
Vt. 05769
Crops: tomatoes, squash

Dr. Arni Hendin 80
RD 1 pt
Randolph Center, Vt. 05061 po
Crops: eggs, hay, apple trees
Quantities: 400 apple trees, 70,000 tons hay
Livestock: 20-60 hogs, goats
Sell harvest to friends.

Raymond Brown 20
Rochester, ft
Vt. 05767 po
Crops: fresh vegetables, dry beans
Livestock: 6 head cattle
I market my harvest through direct sales.

Robert O'Brien 175
Tunbridge, ft
Vt. 05077 ao
Crops: sheep and vegetables (potatoes, carrots, etc.)
Livestock: 30 lambs annually
Sell to mostly old customers.

Burton D. Heath ¼
So. Hill pt
Williamstown, Vt. 05679 ao
Crops: corn, squash, beans, tomatoes
Quantities: 6 bu.

Olin Tewksbury 10
30 Bridge St. pt
Windsor, Vt. 05089 ao
Crops: all common vegetables
Livestock: 4 herefords
Roadside stand

W. L. Brenneman 200
Fiddlehead Farm pt
Worcester, Vt. 05682 ao
Crops: vegetables, milk, eggs, lamb
Livestock: goats, chickens, sheep
Direct to retail stores and at house
Regular commercial channels

VIRGINIA

John B. Parker, Jr. 33
6219 Villa St. pt
Alexandria, Va. 22310 po
Crops: corn, pumpkins, tomatoes, fruits
Quantities: 10 tons corn, 12 tons pumpkins, 1 ton tomatoes, 4 tons apples, grapes, nuts
Livestock: 40 hogs

Mr. & Mrs. Milo Sonen 30
6021 N. 16th St. pt
Arlington, Va. 22205 po
Crops: vegetable garden
Livestock: 10 angus steers
I sell to my regular customers.

T. R. Barker 4
Wisharts Point, Box 189 ft
Atlantic, Va. 23303 ao
Crops: various vegetables
Quantities: 20 bushels
Roadside stand
Regular commercial channels

Harold G. Kruse
Hickory Hill Farm ft
Loganville, Wisc. 53943 ao
Crops: corn, oats, hay, sorghum, mixed vegetables
Quantities: 25 acres, 20 acres, 60 acres, 1 acre and 3 acres mixed vegetables
Livestock: cattle, chickens, ducks, geese, 35 milk cows, 20 steers, 200 hens, 200 roosters, 150 ducks
Roadside stand, direct to retail store and organic shopping directory customers.
Regular commercial channels

Richard M. Higgs 61
RR #1 pt
Merrimac, Wisc. 53561 po
Crops: corn and vegetables
Livestock: cattle, fowl, pigs, rabbits

Marum Marty 225
R. 1 ft
Monticello, Wisc. 53570 ao
Crops: corn, oats, alfalfa
Livestock: 55 cows, 30 beef cattle, 70 head heifers
Regular commercial channels

Josef Part 25
Rt. 4 pt
Mosinee, Wisc. 54455 ao
Crops: oats, hay, vegetables, beef
Quantities: 500 bu. oats
Livestock: 1 heifer, 6 calves

Arthur Klingbyll 2
Rt. 2, Box 182 pt
Mukwonago, Wisc. 53149 ao
Crops: berries, melons, tomatoes
Quantities: ½, ¼, ¼, approx. acres
Roadside stand and the plant where I work.

Serge Ledwith 80
Rt. 3 ft
Muscoda, Wisc. 53735 ao
Soil building-organic conversion
Livestock: 30 milk cows

Gilbert J. Hein 60
R.R. 1, Box 121 pt
Neosho, Wisc. 53059 ao
Crops: hay, oats, corn
Livestock: beef, pork, chickens
Regular commercial channels

Albert R. Popp 65
Rt. 2 pt
New Holstein, Wisc. 53061 ao
Crops: hay, corn, oats, garden vegetables
Livestock: 15 cattle, 20 pigs, 150 chickens, 50 ducks, 4 ponies

Harvey Considine 200
Diamond Dairy Goat Farm ft
P.O. Box 133 ao
North Prairie, Wisc. 53153
Crops: alfalfa, corn, oats, produce cheese, goat milk
Quantities: hay 400 tons, corn 2000 bu., oats 1500 bu.
Livestock: 600 purebred dairy goats
Wholesaler, direct to retail store and direct to consumer.
Regular commercial channels

William Bennett 1
Rt. 2 pt
Oregon, Wisc. 53575 ao
Crops: garden (family) surplus for sale
Roadside stand and direct to retail store

William A. Maas 20
330 River St. ft
Portage, Wisc. 53901 ao
Crops: corn, oats, vegetables
Livestock: chickens, ducks, rabbits, goats, and steers

William Steil 25
Rt. 1 pt
Potosi, Wisc. 53820 ao
Crops: corn & vegetables
Livestock: 4 pigs, 2 calves and 8 chickens

Bro. S. J. Staber 39
411 Parish St. pt
Prairie DuChien, Wisc. 53821
Crops: hay
Quantities: 3,000 bales

Herman J. Haack ¼
2101 Dekoven pt
Racine, Wisc. 53403 po
Crops: strawberries, vegetables and other fruits
Livestock: rabbits
Direct to retail store

Mrs. Marcella Westphal 30
Rt. 2 ft
Reedsburg, Wisc. 53959 ao
Crops: corn, alfalfa, sorghum
Livestock: 15 milk cows, 10 head of young stock

Robert J. Rothstein 5
Rt. 1 pt
River Falls, Wisc. 54022 ao
Crops: tomatoes, sweet corn
Quantities: small
Livestock: chickens, geese, ducks, goats
Roadside stand

S. A. Isaacson 60
Scandinavia, pt
Wisc. 54977 ao
Crops: potatoes, squash, berries, carrots, beets, corn, onions, oats, hay
Quantities: 150 bu. potatoes
Livestock: 2 cows, 100 chickens
Roadside stand
Regular commercial channels

C. B. Lindsey 60
P.O. Box 7 ft
Siren, Wisc. 54872 ao
Crops: carrots, all root crops, also smaller amounts of all vegetables and fruits
Livestock: chickens, rabbits, eggs & goat milk
Direct to retail store
Market through customer advertising and organic guide.

Jerry Tesser 25
1201 S. Duluth Ave. pt
Sturgeon Bay, Wisc. 54235 po
Crops: hay, corn for cattle

William Krug, Jr.
W239 N7332 Maple Ave. pt
Sussex, Wisc. 53089
Just purchased 38 acres will take a couple of years to get started.
Crops: fruits & veg., strawberries, asparagus, tomatoes

Emmet Thorland 51
Rt. 3 pt
Watertown, Wisc. 53094 po
Crops: corn and hay
Quantities: 1100 bu. corn and 20 tons hay
Livestock· hogs and beef, chickens and guineas
Regular commercial channels

Lyn O. Heise 160
Rt. 3 ft
Wausau, Wisc. 54401 po
Crop: ginseng
Quantities: 2000 lbs. per year
Livestock: cattle, pigs, and chickens
Regular commercial channels plus direct mail.

Ronald Zettler 25
R.R. 6 pt
West Bend, Wisc. 53095 ao
Crops: hay, oats
Livestock: goats, rabbits, chickens and ducks
Regular commercial channels

CANADA

Ken Ashworth 700
Deadwood, Alberta, ft
Canada ao
Crops: wheat, oats, barley, rye and alfalfa
Quantities: 11,000 bushels grains
Livestock: 200 hogs and 4 cattle
Regular commercial channels

Roland Gillis 50
Box 744 pt
Castlegar, B.C., Canada po
Crops: fruits and vegetables
Hasn't been farmed for 17 years.

Mr. & Mrs. W. F. Pitt 35
Seafield Farm pt
La Fortune Rd., R.R. 1 ao
Cobble Hill, B.C., Canada
Crops: hay, lamb, apples, veal
Quantities: 50 tons hay
Livestock: dairy cattle, sheep
Sell to friends

Mrs. J. E. Thompson
2-1809 Crescent Rd. ft
Victoria, B.C., Canada ao
Crops: vegetable, dwarf fruits, berries and
honey
Roadside stand

Miss Jutta Mueller 75
R.R. 2, Burks Falls ft
Ontario, Canada ao
Crops: hay and pasture
Livestock: beef cattle, rabbits, chickens, geese
Regular commercial channels

Sammy Szabo 150
Box 24, Gore Bay ft
Ontario, Canada

Peter Nippo 120
P.O. Box 171 pt
Lucan, Ontario, Canada ao
Crops: beef and hogs, white beans, soybeans,
buckwheat
Livestock: 20 beef and 30 hogs
Regular commercial channels

John W. Sigsworth 20-25
300 Hamber Ave. pt
Oshawa, Ont., Canada po
Crops: hay, beef cattle, vegetables
Quantities: 1000 bales hay
Livestock: 7 beef animals
Regular commercial channels

Edwin Alexander 80
R.R. 4 ft
Picton, Ontario, Canada po
Crops: peas, barley, oats, hay
Quantities: 20 acres peas, 260 bu. barley, 260
bu. oats

Roddie MacEachern 30
R.R. 1 pt
Prizeville, Ont., Canada po
Crops: pasture
Livestock: 2 horses

Calvin Gingerich 86
R.R. 2 ft
Zurich, Ont., Canada ao
Crops: oats, barley, wheat, soybeans
Quantities: 3800 bu.
Livestock: cows, pigs, roosters
Regular commercial channels

Mrs. Donald MacQuarrie 500
Cardigan P.O. ft
Roseneath, P.F.I., Canada po
Crops: potatoes, grain and clovers
Livestock: beef cattle
Wholesaler

L. G. Levie 60
R.R. 1 ft
St. Chrusostome, ao
Co. Chateauguaym, Prov.
Quebec, Canada
Crops: vegetables, maple syrup
Livestock: beef cattle
Sell my harvest to friends.

George Thornton 40
Box 282 ft
Bengough, Sask., Canada po
Crops: corn, potatoes, cucumbers
Livestock: chickens, pigs, goat
Direct to retail store and customers.

A. Scheresky 1280
Box 10 ft
Glen Ewen, Sask., Canada po
Crops: wheat, flax, millet
Quantities: wheat 8000 bu., flax 4000 bu., and
millet 1000 bu.
Sell direct to customer.
Regular commercial channels

FOREIGN COUNTRIES

Errol L. Johnstad 160
Manfred-von-Richthofen-Str. 34
1 Berlin 42, Germany
Expect to start farming in 1971.

Dave Theobold 50
Wilderland, RD 1 ft
Whitranga, New Zealand ao
Crops: salad vegetables and fruit
Livestock: 5 goats
Roadside stand

114

Organic Farm Survey

Organic Farmer Survey
Organic Gardening and Farming
Emmaus, Pa. 18049

I now farm .... acres of tillable land.

I farm it on a ( ) full-time   ( ) part-time basis.

I now operate my farm on an ( ) all-organic basis.
                             ( ) partly organic basis.

My main crops are ....

Approximate quantities produced are ....

Livestock include ....

I sell my harvest to the organic market through a

( ) wholesaler   ( ) direct to retail store

( ) roadside stand   ( ) other (please specify)

I market my harvest through

( ) regular commercial channels   ( ) others

(TEAR OUT AND MAIL)

CHAPTER 25

Organic Fertilizer Directory

This directory is a service without charge to those listed. The purpose is to provide as comprehensive and accurate a listing as possible of sources of organic fertilizers and soil conditioners, mulches and mulching materials for the organic gardener or farmer. The editors aren't able to investigate each listing and must rely on the sincerity of those offering farming materials for the organic market.

ARIZONA

Chemor Products Co.
Phoenix 85000
Peat moss & organic mulches.

ARKANSAS

Natural Resources Development
Clidot, Inc. Soil Conditioners
Grannis 71944
Phone: (501) 385-2439

CALIFORNIA

Ecology Trading Center
788 Old Country Road
Belmont 94002
Phone: (415) 592-0305
San Francisco

Eden Acres Farm
Organic Compost-Fruit
1264 E. Alvarado Street
Fallbrook 92028

Fred S. Elesh, M.D.
12412 9th Street
Garden Grove 92640
Compost containing kelp meal, bone meal, cottonseed meal, fish meal and ground oyster shell.
Retail.

Forci-Grow
Rt. 1, Box 1866
Lathrop 95330
Organic fertilizers & soil conditioners.

Gourmet Mushroom Farms
100% Organic weedfree COMPOST
6036 American Ave.
Modesto 95350
Phone: (209) 529-8458

Pace-Co/General Organics
Box 8082 (O)
Oakland 94608
Phone: 533-0549 (SF: 585-1016)
Also southern California outlet.
Retail, wholesale ship out of state & take mail orders.

Wright Feeds
16210 South Colorado
Pine Center
Paramount 90723
Sea-gro.
Fish & sea-derived fertilizers.

The Fersolin Corporation
100 Bush Street
San Francisco 94104
Lomite.
Organic fertilizers & soil conditioners.

Organic Farm & Garden Center
924 Olmstead Street
San Francisco 94134
Fertilizer.

Ocean Labs, Inc.
Berth 42-Outer Harbor
San Pedro 90731
Phone: (213) 832-7274
Kelp.

Robert Lefever
c/o AFCO Nursery
204 N. Blosser Road
Santa Maria 93454
Natural fertilizers.

Saddleback Soil Service
Producer of Unisoil compost
29361 Via Portola
South Laguana 92677

Vita Green Seed & Fertilizer Co.
P.O. Box 878
Vista 92083
Phone: (714) 724-2163

Rich Davis
3263 Sugarberry Lane
Walnut Creek 94598

COLORADO

U-Need-Us Fertilizer Co.
P.O. Box 26
Austin 81410
Phone: 835-5335
Trace mineral fertilizer.

116

Kitner Health Food
Kelp seaweed and concentrate.
4 miles north of
Cedaredge 81413

Rich-Loam
Colorado Springs 80900
Organic fertilizers & soil conditioners.

Bruce L. Yarbrough
Lazy-Glen Ranch & Greenhouse
Snowmass 81654
Natural fertilizers.

CONNECTICUT

Brookside Nurseries, Inc.
228 Brookside Road
Darien 06820
Phone: (203) 655-3978
    (203) 853-2076 (Norwalk)
Sell retail, wholesale, make out-of-state shipments.
Organic fertilizers & soil conditioners.

Durham Fertilawn & Soil Builders
Cherry Land
Durham 06422
Natural fertilizers.

Mineral Division of Skod Co.
10 Lewis Street
Greenwich 06830
Seaborn.
Fish & sea-derived fertilizers.

United Rent-alls
328 Paddock Avenue
Meriden 06450
Mr. E. Salamandra
Natural fertilizers.

Green Warehouse Co.
34 Bridge Street
New Milford 06776
Phone: 354-4285
Fertilizer.

Charles M. Perol
Elmbrook Farm, Torrington Road
Winsted 06098
Fertilizer.

Pine Willow Farms
North Taylor Avenue
So. Norwalk 06854
Natural fertilizers.

DISTRICT OF COLUMBIA

Cosmic View, Inc.
Greensand, granite dust, phos., rock, etc.
4822 Mac Arthur Blvd. NW
Washington 20007
Phone: (202) 333-1737

FLORIDA

Reid Fertrell Distributing
123 14th Avenue S.W.
Largo 33540
Natural fertilizers.

Lee's Fruit Company
Box 450
Leesburg 32748
5 mi. N on U.S. Rest. 27 & 441.
Rock phosphate, granite meal, worm soil, blood meal, Tankage, cotton seed meal, micro min, compost.
Sell retail, make out-of-state shipments.

Carlton W. Neier
6730 South Drive
Melbourne 32901
Phone: (305) 723-3328
Colloidal phosphate, compost, sludge, cotton seed meal, tobacco stems, etc. Sell retail.

Atlantic & Pacific Research Inc.
Box 14366
North Palm Beach 33403
Sea-magic.
Fish & sea-derived fertilizers.

Mr. O. K. Mairs
Box 7413
College Park Station
Orlando 32804
Natural fertilizers.

Sunshine Poultry Farm
N. Lockwoodridge Road
Sarasota 33580
Fertilizer.

Robert Haber
Tall Organics Co.
Box 1326
Tallahassee 32302

James B. Conklin
701 West Highway 54
Zephyrhills 33599
Phone: 782-1859
Wholesale, out-of-state shipments.

GEORGIA

Hybro-Tite Corp.
Box 527
Lithonia 30058
Phone: (404) 482-7346
Sell retail, wholesale, out-of-state shipments.
Organic fertilizers & soil conditioners. Hybro-Tite

Blenders, Inc.
Joe S. Francis
Lithonia 30058

John Farmer
274 South Four Lane
Marietta 30060
Natural fertilizers.

ILLINOIS

Lake-Cook Farm Supply & Garden Stores
510 E. Northwest Highway
Arlington Heights 60804
Phone: 253-0570
Sell retail & wholesale.

Burlington Sales Co.
433 New York Street
Aurora 60505
Natural fertilizers.

Martin Stansbury
1608 Franklin
Bloomington 61701
Natural fertilizers.

Wonder Life Company
Henry Rothermel
Broadlands 61816
Phone: (217) 834-3385

Home & Garden Supply Co.
John De Koker
4701 West 55th Street
Chicago 60632
Phone: PO 7-1200
Sell retail, make out-of-state shipments.

Irving Park Dietary Foods
3931 W. Irving Park
Chicago 60618
Phone: KE 9-1490
Sell retail.

Oil-Dri Corp. of America
520 N. Michigan Avenue
Chicago 60611
Phone: (312) 321-1515
Sell wholesale, make out-of-state shipments.
Organic fertilizers & soil conditioners.
Terra-Green.

Sea-Born Corporation
3421 N. Central Avenue
Chicago 60634
Algit Norwegian kelp meal, feed supplement,
Sea-born liquid and granular seaweed for soil
use.
Sell retail, wholesale, make out-of-state ship-
ments.

The Zonolite Company
135 S. La Salle Street
Chicago 60603
Terra-Lite.
Organic fertilizers & soil conditioners.

Robert E. Schwaan
Schwaan's Garden Center
5324 W. 26th Street
Cicero 60650
Fertilizer.

Lake-Cook Farm Supply & Garden Stores
997 Lee Street
Des Plaines 60016
Phone: 824-4406
Sell retail & wholesale.

Klueter Feed Store, Inc.
Rt. 159
Edwardsville 62025
Natural fertilizers.

Lake-Cook Farm Supply & Garden Stores
381 Center St.
Grayslake 60030
Phone: 223-2344
Sell retail & wholesale. Fertilizer.

Midwest Phosphate Co.
Essington Road
Joliet 60435
Organic fertilizers & soil conditioners.

Blue Ribbon Iris Gardens
William Macuotka
9717 W. 55th St.
Countryside P.O.
La Grange 60525
Phone: FL 4-2966

Wm. V. Machotka
Blue Ribbon Iris Gardens
9717 W. 55th Street
La Grange 60525
Phone: FL 4-2966
Sell retail. Natural fertilizers.

Lake-Cook Farm Supply & Garden Stores
Railway Avenue
Lake Zurich
Phone: 438-2161
Sell retail & wholesale.

Soil Life Plant Food
Soil Life Research Co.
Division of Ray Rees & Sons Inc.
PawPaw 61353

Lake-Cook Farm Supply & Garden Stores
9 So. Roselle Road
Roselle 60172
Phone: 529-3601
Sell retail & wholesale.

Sunnylawn Farm
Hybro-tite dist. Korn-Cob mulch
Box 101
Steward 60553

Lake-Cook Farm Supply & Garden Stores
6730 South Street
Tinley Park 60477
Phone: 532-4723
Sell retail & wholesale.

Lake-Cook Farm Supply & Garden Stores
3469 N. Sheridan Road
Zion 60099
Phone: 872-8200
Sell retail & wholesale.

INDIANA

Benson-Maclean
Bridgeton 47836
Natural fertilizers.

Tyner's Organic Park
R.D. #2
Stevenson Station Road
Chandler 47610
Phone: 867-5050
Natural fertilizers and mulches. Sell retail.

Heniz Mfg. Co., Inc.
Farm & Fairway Div.
P.O. Box 318
Elwood 46036
Natural fertilizers.

Associated Specialists
P.O. Box 21314
Indianapolis 46221
Phone: (317) 241-3866
Compost, also sand, gravel. Wholesale only.
Natural fertilizers.

Norwegian Sea Kelp Prods.
2223 Lafayette Road
Indianapolis 46222
Phone: (317) 638-2081
Retail, wholesale, make out-of-state shipments.
Natural fertilizers.

The Gardener's Cupboard
P.O. Box 61
12 Points Station
Terre Haute 47808
Natural fertilizers.

IOWA

Hy-Brid Sales Company
924-26 South 6th Street
P.O. Box 276
Council Bluffs 51501
Phone: (712) 323-5022
Sell retail, wholesale, make out-of-state ship-
ments. Natural fertilizers.

Kimberly Barn, Re-Vita
Mineralizer soft rock phos., granite meal, lime
1221 East Kimberly Road
Davenport 52807

Farmcraft Soiltrate
Box 1431
Des Moines 50305
Natural fertilizers.

L. W. Kauten
Fayette 52142
Phone: (319) 425-4088
Natural fertilizers & soil conditioners.
Sell retail, wholesale to distributors.

Norwegian Sea-Weed Products
Laurence Beckert
206 Ave. D
Fort Madison 52627

F. S. Brisbois
Fonda 50540
Phone: (712) 288-6578
Sell retail, wholesale, make out-of-state ship-
ments.
Kelp meal.
Fish & sea-derived fertilizers.

Harry S. Halvorsen
Rt. #3
Forest City 50436
Phone: 582-4673
Sell retail, wholesale. Organic fertilizers.

Leslie L. Byler
Kalona 52247
Natural fertilizers.

KANSAS

Bob & Dorothy Arnett
Arnett Natural Food & Farming Supplies
836 S. Summit
Arkansas City 67005

KENTUCKY

Clair W. Stille
130 N. Hanover Ave.
Lexington 40502
Phone: (606) 266-8066
Fertosan organic compost maker. 100% organic liquid compost concentrate.
Sell retail, wholesale, make out-of-state shipments.

Happy Acres, Inc.
Miracle Soil Tone Mfrs.
P.O. Box 711 (on Hwy. 1247, Elihu)
Somerset 42501
Phone: (606) 678-8896
Natural soil conditioners, potash, iron, minerals. Sell retail, wholesale, make out-of-state shipments.

MARYLAND

Wengers Farm & Garden Center
Box 80, Route #1
Mechanicsville 20659
Natural fertilizers.

Eli Yoder
Route 2, Box 129
Oakland 21550
Phone: (301) 334-9254
Sell retail, wholesale, some out-of-state shipments.
Organic soil conditioners, fertilizers.

Maryland Bay Soil & Fertilizer Co.
Shady Side 20867
Organic fertilizers & soil conditioners.

MASSACHUSETTS

Emanuel & Jeanne Occhipinti
Plant — Tone distributor
Shady Lane Greenhouse
Rt. 140 & 31
E. Princeton 01517

Bruckmann's
Fetrell, Rock phos., cottonseed, etc.
79 (rear) S. Broadway
Lawrence 01843

Alexander's Blueberry Nurseries
R.F.D. 4, Box 299
1224 Warcham Street
Middleboro 02346
Phone: (617) 947-3397
Sell retail, make out-of-state shipments (not fertilizer.)

J. Herbert Alexander
Dahliatown Nurseries
Middletown 01949
Organic fertilizers & soil conditioners.

Willard C. Rutherford
Squanto Peat & Organic Fertilizer Co.
Oakham 01068
Phone: 882-5279
Sell retail, wholesale, make out-of-state shipments.
Soil conditioners, fertilizers, mineral products, rock phosphate, granite dust, cotton seed meal, bone meal, dried blood, fish fertilizer, organic insecticide.

Pentti M. Penttila
197 West Street
Paxton 01612
Natural fertilizers.

Odlin H. Day
Yard & Garden Specialties
753 Dennison Drive
Southbridge 01550
Natural fertilizers.

Conrad Fafard, Inc.
P.O. Box 2131
Springfield 01101
Phone: (418) 739-2101
Sell wholesale, make out-of-state shipments.
Peat moss & organic mulches.

Organic Sure Gro Mfgr.
Triple A Mills
A. Aksila, Manager
125 Pierce Road
Townsend 01469
Phone: 597-2063

The Garden N' Gift Shop
35 North Main Street
On Old Cape Rt. 28
West Bridgewater 02379
Natural fertilizers.

MICHIGAN

Dick A. Dean
Box 517 Tudor Rd.
Berrien Springs 49103
Phone: (616) 477-3765
Sell retail. Natural fertilizers.

Granzow's Organic Garden
Organic fertilizers.
8350 Dixie Highway
Clarkston 48016

Allen's Lawn & Garden Store
2925 Francis Street
Jackson 49203
Phone: (517) 789-8927
Sell retail, wholesale, make out-of-state shipments.
Natural fertilizers.

Fanning Soil Service Inc.
Jess M. Fanning, President
4951 S. Custer Road
Monroe 48161
Phone: 241-7570
Sell retail, wholesale, make out-of-state shipment.
Natural soil minerals, colloidal phosphate.

How's Dairy Goat Farm
1406 North Monroe Street
Monroe 48161
Natural fertilizers.

Saxton's Garden Center Inc.
586 W. Ann Arbor Trail
Plymouth 48170
Phone: (313) 453-6250
Sell retail, wholesale, make-out-of-state shipments.
Natural fertilizers.

Telezinski Natural Organic Fertilizer
17230 Twelve Mile Road
Roseville 48066
Phone: PR 6-4619
Sell retail. Natural fertilizers.

Duncan & Sons
1102 South Washington
Royal Oak 48067
Natural fertilizers.

Theo T. Juengel
11 E. Grove St.
Sebewaing 48759
Phone: (517) 881-3241
Sell retail, wholesale, make out-of-state shipments.

Thomas L. Telford
"Uncle Luke's Feed Store"
Fertrell Distributor
6691 Livernois
Troy 48084

MINNESOTA

Perkins Crosslake Garden Center &
Nursery, Inc.
Crosslake 56442
Phone: (218) 692-3805
Sell, retail, few out-of-state shipments.

Farmer Seed & Nursery Co.
Faribault 55021
Natural fertilizers.

Delmer C. Bunke
Canton Mills, Inc.
Minnesota City 55959

Soil Service
Sleepy Eye 56085
Natural fertilizers.

Super Gro Products Co.
Box 86
Winona 55987
Phone: 452-2112
Natural fertilizers.

Winona Farm & Garden Center
116 Walnut Street
Winona 55987
Natural fertilizers.

MISSISSIPPI

Organic Green-Grow Service
P.O. Box 342
Crystal Springs 39059
Natural fertilizers.

MISSOURI

O. K. Hatchery
Feed & Garden Store
140 E. Madison Avenue
Kirkwood 63122
Phone: 822-0083

NEBRASKA

Mr. Glayne D. Doolittle
Route #8
Lincoln 68501
Natural fertilizers.

Organic Soil Conditioner Prod.
Route #8
Lincoln 68501
Natural fertilizers.

NEW JERSEY

Far Hills Nursery, Inc.
Robert S. Ratti
U.S. 206
Bedminster 07921
Phone: (201) 234-0906
Sell retail.

Organic Growth Prod. Co.
P.O. Box 446
Farmingdale 07727
Natural fertilizers.

Swartzel's Farm & Garden Center
645 Holmdel Road by RR
Hazlet 07730
Phone: 264-2211
Organic fertilizers & soil conditioners.
Sell retail & wholesale.

Goulard & Olena, Inc.
Skillman (Somerset County)
Hopewell 08525
Natural fertilizers.

The Esboma Company
Millville 08332
Organic fertilizers & soil conditioners.

New Brunswick Flour Co.
251 Neilson Street
New Brunsick 08901
Natural fertilizers.

The Hyper-Humus Co.
P.O. Box 267
Newton 07860
Phone: (201) 383-2300
Humus, organic fertilizers, soil conditioners.
Sell retail, wholesale, make out-of-state shipments.

Peralex of New Jersey, Inc.
64 Barclay Street
Paterson 07503
Perlite.
Organic fertilizers & soil conditioners.

Quaker Lane Products
Pittstown 18951
Powdered cow manure.
Sell retail, wholesale, make out-of-state shipments.

Parker Greenhouses
Chicken manure
1325 Terrill Road
Scotch Plains 07076
Phone: (201) 322-5552

Orol Ledden
Orol Ledden & Sons
Center & Atlantic Avenue
Sewell 08080

NEW MEXICO

International Humate Producers &
Distributors
701 Madison N.E.
Albuquerque 87110
Phone: (505) 268-9206
Humic acid fertilizers, organic minerals, diatomaceous, earth insecticides.
Sell retail, wholesale, make out-of-state shipments.

Vitabloom
4100 Silver S.E.
Albuquerque 87108
Organic fertilizers, soil conditioners.

Loloma Nursery
Fertex, garden supplies,
vegetable plants, trees.
Rodeo Road
Santa Fe 87591
Phone: (505) 982-3310

NEW YORK

Sealife Products Inc.
Box 454
Amagansett 11930
Fish & sea-derived fertilizers.

Klein's Evergreen Farm
8150 Greiner Road
Buffalo 14221
Natural fertilizers.

Enoch Eichorn
5270 Salt Road
Clarence 14031
Phone: (716) 759-6320
Sell retail, wholesale.

Larrowe Mills Inc.
Cohocton 12726
Mul-Tex.
Peat moss & organic mulches.

Ra-Pid-Gro Corp.
88 Ossian Street
Dansville 14437

Deer Valley Farm
R.D. #1
Guilford 13780
Natural fertilizers.

John H. Miller
5748 Campbell Blvd.
Route #2
Lockport 14094
Natural fertilizers.

Massapequa Seed & Garden Supply
500-504 Hicksville Road
Massapequa, L.I. 11758
Phone: (516) 798-0123
Sell retail, wholesale, make out-of-state shipments.
Organic fertilizers, soil conditioners.

Liffco, Inc.
80 Herricks Road
Mineola, L.I. 11501
Phone: (516) 746-1900
Sell retail, wholesale, make out-of-state business.
Natural fertilizers.

John D. Cramer
10702 Cayuga Drive
Niagara Falls 14304
Natural fertilizers.

Wilson Greenhouse
20 South Creek Road
Pine City 14871
Phone: 734-6350
Sell retail.
Natural fertilizers.

Abbey Organic Gardens
184 McCall Road
Rochester 14616
Natural fertilizers.

East Ridge Garden Store
160 East Ridge Road
Rochester 14621
Natural fertilizers.

General Maintenance Service
65 Fairfax Avenue
Schenectady 12304
Natural fertilizers.

Almith Industries
Box 275
Seaford 11783
Sell retail, wholesale, make out-of-state shipments.
Organic fertilizers & soil conditioners.

Threefold Farm Dairy
Hungry Hollow Road
Spring Valley 10977
Natural fertilizers.

Valley Feed & Supply Co., Inc.
Union Road & Railroad
Spring Valley 10977
Natural fertilizers.

Sterling Forest Peat Company
Box 608
Tuxedo 10987
Phone: (914) 351-2246
Sell retail, wholesale, make out-of-state shipments.
Peat moss & processed organic soil.

Gardener's Village
Hempstead Tpk. & Cherry Valley Rd.
(opp. Island Garden Arena)
West Hempstead, L.I. 11552

Organic Supply Co.
15 Hawthorne Street
White Plains 10603
Natural fertilizers.

OHIO

Troyer Fertrell Dist.
Baltic
Natural fertilizers.

Hiffner's Organic Gardens
Re-Vita mineralizer, Hybro-Tite,
Everything for organic growers
2450 Hazel Drive
Beavercreek
Xenia 45385

Eisman Organic Garden Supplies
6046 Benken Lane
Cincinnati 45211
Phone: 661-4944
We have everything to make your garden grow organically.
Sell retail, wholesale, make out-of-state shipments.

McDonald's Organic Supplies
Fortrel Fertilizer, complete line
8968 Kenwood Road
Cincinnati 45242
Phone: 791-8480

Nature's Way Products
Fred A. Veith
3505 Mozart Avenue
Cincinnati 45211
Phone: (513) 481-0982
Rock powders and organic nitrogens.

Wagoner Bros.
Route 1, Union Road
Clayton
Natural fertilizers.

FTE
The Ferro Corporation
Cleveland
Phone: (216) 641-8580
(fritted trace elements)
Sell wholesale, make out-of-state shipments.

Jack S. Alkire
42 S. Burgess Avenue
Columbus
Natural fertilizers.

Beechwold Natural Foods
4185 North High Street
Columbus
Natural fertilizers.

General Products of Ohio
Crestline
Compost and dehydrated manure.

Mr. Jack Hetzel
Paygro, Inc.
P.O. Box 768
Dayton
Natural fertilizers.

Ree-Tern Inc.
National Distribution
5335 Far Hills Avenue
Dayton 45429

L. Henkhuzens—L Vegetate
744 E. 203rd Pl.
Euclid 44119

Paul Frech
Box 88, R.R. #1
New Paris 45347
Natural fertilizers.

North Olmsted Feed & Supply
27100 Lorain Road
North Olmsted 44070
Natural fertilizers.

Roto-Hoe Sales & Service
6621 Troy Road, Rt. 2
Springfield 45502
Phone: 964-1030
Sell retail.
Fertilizers, rock phosphate.

Vessey's Nursery & Garden Center
477 Northeast Ave., Rte. #261
Tallmadge 44278
Phone: 633-3627

Dietrich's Fertrell Dist.
519 Monroe Street
Toledo 43604
Natural fertilizers.

Norbert J. Heckman
Route 705
Yorkshire 45388
Natural fertilizers.

OKLAHOMA

"Gro-Crop" Dolomite
Delta Mining Corporation
P.O. Box 95
Mill Creek 74856
Phone: (405) 384-2500

Natural Plant Food Co.
1409 N.W. 50th Street
Oklahoma City 73118
Sell wholesale, make out-of-state shipments.
Dehydrated manure.

U. S. Organic Ferto Co.
100% pure organic fertilizer
1133-B S. Memorial Drive
Tulsa 74112
Phone: (918) 836-5106

OREGON

Ore-gano
Organic plant food & trace minerals
Oregon Organic Gardens
Box K
Cave Junction 97523

PENNSYLVANIA

Natural Development Co.
John A. Johnson
Box 215
Bainbridge 17502
Phone: (717) 367-1566
Fertilizers & soil conditioners.
Sell retail, wholesale, make out-of-state shipments.

Brook's Builders Supplies
2520 8th Avenue
Beaver Falls 15010
Natural fertilizers.
Phone: (412) 846-7584

Musser's Fertrell Dist.
Bowmansville 17507
Natural fertilizers.

Harold Anderson
8 Levi Street
Warren Co.
Clarendon 16313
Natural fertilizers.

Krear Service Station
Rt. 38
Emlenton 16373
Phone: (412) 867-2220
Fertrell, greensand, rock phosphate, & tri-excel-DS.

James P. Webb
Webb Super Gro Products
Webb Building
Flemington 17745
Natural fertilizers.

Zook & Ranck, Inc.
Floyd H. Ranck
R. #1
Gap 17527
Phone: (717) 442-4171
Compost, natural fertilizers.
Sell retail, wholesale, make out-of-state shipments.

Best Feeds & Farm Supplies, Inc.
Rt. 8—1656 Wm. Flynn Highway
Glenshaw 15116

James D. Miller
2147 Mt. Carmel Avenue
Glenside 19038
Natural fertilizers.

Stoudt's Garden & Yard Service
Harold R. Stoudt
Route 1, Box 557
Hamburg 19526
Phone: (215) 562-8766
Natural fertilizers.
Sell retail.

Hershey Estates
Hershey 17033
Phone: (717) 533-9460
Mulch, peat moss.
Sell wholesale, make out-of-state shipments.

Don Leap
Tri-excell DS, rock phos.
Hyndman 15545
Phone: (814) 842-3370
Rock phosphate, fertilizer.
Sell retail, make out-of-state shipments.

Natural Soil Builders
Brook Lawn Farm
118 Kreider Avenue
Lancaster 17601
Sell retail, wholesale.
Natural fertilizers.

H. Mervin McMichael
R.D. #6, Box 63
Lancaster 17604
Natural fertilizers.

Mr. Kurt Z. Cockley
Box 327, R.D. #1
Landisburg 17040
Natural fertilizers.

Woodcrest Mink Ranch
Elwood Brunt
R.D. #2
Lansdale 19446
Natural fertilizers.

Hofstetter's Organic Products
Nybro-Tite, Fertrell, rock phos., various meals
Limeport, Lehigh County 18060
Phone: (215) 967-2026

Oliver S. Brock
R.D. #4
Meadville 16335
Sell retail, wholesale.
Natural fertilizers.

John W. Neff
Route #2
Mt. Joy 17552
Natural fertilizers.

Francis H. Jarrard
R.D. #1
Nescopeck 18635
Natural fertilizers.

L. W. McConnell
202 S. Crawford Ave.
New Castle 16101
Natural fertilizers.

Aquetong Automotive
Edward Batterson
3 Miles South on Rt. #202
New Hope 18938
Phone: (215) 862-2149

Lund's Store
R.D. #1
New Stanton 15672
(near Arona)
Natural fertilizers.

Best Feeds & Farm Supplies, Inc.
Oakdale 15071

Organic Compost Corp. of Pa.
Box 70
Oxford 19363
Phone: (215) 932-2396
Natural fertilizers.
Sell wholesale, make out-of-state shipments.

Paul Keene
Walnut Acres
Penns Creek 17862
Organic fertilizers & soil conditioners.
Sell retail, make out-of-state shipments.

Best Feeds & Farm Supplies, Inc.
280 Corliss Street
Pittsburgh 15220
Phone: 711-3551

Best Feeds & Farm Supplies, Inc.
7381 McKnight Road
Pittsburgh 15220

Locust Farms
R.D. #1
Prospect 16052
Natural fertilizers.

Huber Bros.
Schaefferstown 17088
Natural fertilizers.

Brabon Research Farm
Telford 18969
Phone: 257-9552
Organic compost.
Sell retail, wholesale, make out-of-state shipments.

H. C. Webber
R.D. #1
Telford 18969
Natural fertilizers.

Best Feeds & Farm Supplies, Inc.
965 Airbrake Avenue
Turtle Creek 15145

SOUTH CAROLINA

U. S. Peat Company
P.O. Box 568
Walterboro 29488

TENNESSEE

Oliver M. Babcock, Jr.
Ruhm Phosphate & Chemical Co.
Box 361
Columbia 38401
Phone: (615) 388-4317
Ground rock phosphate, fertilizers, soil conditioners.
Only wholesale, make out-of-state shipments.

Robin Jones Phosphate Co.
204 23rd Avenue
Nashville 37206
Phone: (615) 254-3486
Sell retail, wholesale, make out-of-state shipments.
Organic fertilizers, soil conditioners.

TEXAS

Bankhead Feed Store
111 South Mesquite Street
Arlington 76010
Natural fertilizers.

The Garden Mart
5108 Bissonnet Street
Bellaire 77401
Phone: (713) 665-6481
Fertilizers, mulches, natural minerals ground.

Organic Compost Corp. of Texas
Fort Worth 76101
Phone: (817) 626-3773
Sell wholesale, make out-of-state shipments.

Clod Buster Distributing Co.
P.O. Box 8764
Houston 77009
Phone: 526-4757
Humus soil conditioners.

Minores Incorporated
4614 Sinclair Road
San Antonio 78222
Phone: (512) 648-3131
Natural fertilizer rich in phosphate, humus and desirable trace elements. (bat guano)

UTAH

Key Minerals Corporation
A. Z. Richards, Jr.
P.O. Box 2364
Salt Lake City 84110
Phone: (801) 467-8522
Sell retail, wholesale, mail orders.

VERMONT

Justin Brande
Box 61, R.D. #1
Middlebury 05753
Natural fertilizers.

Donald G. Davidson
South Royalton
Windsor County 05001
Phone: 763-8189
Organic fertilizers, mulches, soil conditioners.
Sell retail, wholesale, make out-of-state shipments.

VIRGINIA

The Driconure Company
Box 892
Harrisonburg 22801
Organic fertilizers & soil conditioners.

Gregory's Three-Way Mulch
Gregory General Farms
Java 24565
Sell retail, wholesale, make out-of-state shipments.
Peat moss & organic mulches.

T. J. Kirkup
Standard Products Co., Inc.
Kilmarnock 22482
Phone: (703) 435-1633
Sell retail, wholesale, make out-of-state shipments.
Fish emulsion. Sea-derived fertilizers.

Kenneth E. Gorman
Griffin Brothers Inc.
2043 Church Street
Norfolk 23504

Acme Peat Products LTD.
687 No. 7 Road, Dept. 8
R.R. 2
Richmond 23219
Fish & sea-derived fertilizers.

Tangier Sea Organism Co.
Sinclair's Beach
Tangier 23440
Seaweed.
Sell retail, wholesale, make out-of-town shipments.

Greenlife Products Co.
West Point 23181
Phone: (703) 843-5451
Pine bark mulch, soil conditioners, based fertilizers.
Sell wholesale, make out-of-state shipments.

WASHINGTON

Venables Evergreen Gardens
13228 Lake Road
Alderwood Manor
Box 1081
Lynwood
Natural fertilizers.

Bardens Gardens
W-10th Ave. on N.E. 194 St.
Rt. 3 512
Ridgefield 98642
Retail. Makes out-of-town shipments.

Alaska Fish Fertilizer
84 Seneca Street
Seattle 98101
Phone: (206) MU 2-8830
Sell retail, wholesale, make out-of-state shipments.
Fish & sea-derived fertilizers.

Marine By-Products Co.
3308 Harbor Avenue S.W.
Seattle 98126
Fish & sea-derived fertilizers.

Whiz Fish Products Co.
2000 Alaskan Way
Seattle 98108
Fish & sea-derived fertilizers.

Dee & Geneva Nickols
Organic Gardens
Route 1, Box 31
Soap Lake 98851
Phone: CH 6-2322
Compost and granite dust.
Retail, wholesale, make out-of-state shipments.

Evergreen Organic Supplies
John Kossian
Rt. 2, Box 839
Sultan 98294
Phone: SY 3-2751
Sell retail, wholesale, make out-of-state shipments.

WISCONSIN

Organic Compost Company
Germantown 53022
Phone: (414) 242-0500
Sell wholesale, make out-of-state shipments.
Organic fertilizers & soil conditioners.

"Fer-Til" Duck Organic Compost
Grove Compost Co., Inc.
P.O. Box 242
Grafton 53024
Phone: (414) 377-7520

Specialty Foods & Feeds Exchange
P.O. Box 26
Juda 53550
Natural fertilizers.

Mosser Lee Company
Millston 54643
Sell retail, wholesale, make out-of-state shipments.
Organic fertilizers, soil conditioners.

Phillip's Nursery Service
1905 North 19th Street
Milwaukee 53205
Natural fertilizers.

Sewerage Commission City of Milwaukee
Box 2079
Milwaukee 53201
Phone: (414) 271-2403
Organic fertilizers & soil conditioners.

W. G. Slugg Seed & Fertilizers, Inc.
3922 W. Villard Avenue
Milwaukee 53209
Phone: 466-4500
Sell retail, wholesale, make out-of-state shipments.

Floyd Buckelew
13 Depot Street
Ripon 54971
Natural fertilizers.

Theo. & Ursula Kirchner
Route #2
Shiocton 54170
Phone: (414) 757-5560
Organic soil builder 50 # bags or ton lots.
Sell retail.

Ruth Wileden
The Garden Spot
W 239 N. 6548 Maple Ave.
Sussex 53089
Phone: 246-3481
Sell retail.
Natural soil conditioners.

CANADA

Alginure Seaweed Pro.
Box 693
Sidney, B.C.
Fish & sea-derived fertilizers.

Mac Donald & Wilson
562 Beatty Street
Vancouver, B.C.
Make out-of-state shipments.
Organic fertilizers & soil conditioners.

Reindeer Organic Company, Ltd.
Victoria, B.C.
Organic fertilizers & soil conditioners.

Joseph H. Scruggs
Chem-Organic Fertilizers Ltd.
921 Boul. Est.
Louiseville, P.Q.
Phone: (819) 228-5811
Organic fertilizers.

Bedford Organic Fertilizer Co. Ltd.
2045 Bishop Street—Suite 8
Montreal, P.Q.
Vitalite
Organic fertilizers & soil conditioners.

Annapolis Valley Peat Moss Co. LTD.
Berwick, Nova Scotia
Sell wholesale, make out-of-state shipments.
Peat moss, organic mulches.

Ted Bergman
R.R. #2
Leamington, Ontario
Liquid seaweed, also dry form.
Organic fertilizers & soil conditioners.
Sell retail, make out-of-state shipments.

FOREIGN COUNTRIES

"Villa-V" Health Spa
"Hojas Del Monte" leaf mold
Box 1228
Cuernavaco